Multiproduct Plants

Edited by Joachim Rauch

Also of interest

Zlokarnik, M.

Scale-up in Chemical Engineering

ISBN 3-527-30266-2

Zlokarnik, M.

Stirring – Theory and Practice

ISBN 3-527-29996-3

Hiltscher, G., Mühlthaler, W., Smits, J.

Industrial Pigging Technology

ISBN 3-527-30635-8

Sundmacher, K., Kienle, A. (eds.)

Reactive Distillation

ISBN 3-527-30573-3

Sanchez Marcano, J. G., Tsotsis, T.

Catalytic Membranes and Membrane Reactors

ISBN 3-527-30277-8

Weissermel, K., Arpe, H.-J.

Industiral Organic Chemistry

ISBN 3-527-30578-5

Büchel, K. H., Moretto, H.-H., Woditsch, P.

Industrial Inorganic Chemistry

ISBN 3-527-29849-5

Multiproduct Plants

Edited by Joachim Rauch

Translated by Karen du Plooy

WILEY-VCH GmbH & Co. KGaA

Editor

Dipl.-Ing. Joachim Rauch
BASF Aktiengesellschaft
ZET/ZT – A15
67056 Ludwigshafen
Germany

Translator

Karen du Plooy
Raiffeisenstraße 13M
55218 Ingelheim
Germany
Tel.: +49-6132-799 855
Fax: +49-6132-799 857
Email: karen.duplooy@t-online.de

Library of Congress Card No.: applied for

British Library Cataloguing-in-Publication Data
A catalogue record for this book is available from the British Library.

**Bibliographic information published
by Die Deutsche Bibliothek**
Die Deutsche Bibliothek lists this publication in the Deutsche Nationalbibliografie; detailed bibliographic data is available in the Internet at <http://dnb.ddb.de>

© 2003 WILEY-VCH Verlag GmbH & Co. KGaA, Weinheim

Printed in the Federal Republic of Germany
Printed on acid-free paper

Typesetting K+V Fotosatz GmbH, Beerfelden
Printing betz-druck gmbh, Darmstadt
Bookbinding Litges & Dopf Buchbinderei GmbH, Heppenheim

ISBN 3-527-29570-4

Contents

Preface

Those contemplating the use of plants suitable for the production of a number of products will soon find that the body of literature in this area is rather limited. Especially questions related to plant design or the technical equipment for such plants will usually remain unanswered, as there are very few contributions from the chemical industry on this subject.

This state of affairs as well as the encouraging response to a lecture on the different designs and technology of multiproduct plants given at the annual meeting of the GVC (Gesellschaft für Verfahrenstechnik und Chemie [Process Engineering and Chemical Engineering Society]) in Aachen in 1994 has motivated the editor of this book to convince colleagues who are recognized experts in their areas to collaborate on a book on multiproduct plants.

That the concepts and definitions used in this area should be right at the beginning of the book was already obvious during our early discussions. The outcome is collected in Chapter 1, which is followed by descriptions of typical applications of multiproduct plants in Chapter 2. In the next chapter, Chapter 3, the different types of multiproduct plants, subdivided from our point of view, are presented. New developments, such as the pipeless plant concept, are also covered.

It will soon become clear to the reader progressing through the book that the authors of the later chapters always refer back to these different plant concepts, that is, the concepts of the different types of multiproduct plants are used as the central reference point of the book. It is therefore recommended that the sections covering the basic concepts (Chapters 1 to 3) are read before the later chapters.

Our approach to the themes related to multiproduct plants is clearly technical, as shown by Chapters 4 to 11. That contributions on plant organization and the management of multiproduct operations are lacking, may be regarded as a shortcoming. At this point, I would like to express my hope that these interesting themes will be taken up at a later stage.

Many of the authors writing on their specific technical areas have found it useful to aid the reader's comprehension by first introducing the area in general, before continuing with the special aspects relating to multiproduct plants.

For illustrating the machinery, apparatus, and piping and connection technology applied in multiproduct plants, examples of well-established technical components available from suppliers of machinery, apparatus, and plant hardware, were thought to be useful, because important developments in this area have resulted particularly from efforts on the supplier side, so that a variety of technical solutions are now on offer. Only a small selection of components and systems is described here, and our choice is therefore not claimed to be comprehensive. The selection does, however, demonstrate the impressive innovation drive of the manufacturers.

From the final chapter, on the choice and optimization of multiproduct plants, it is clear that many multiproduct plants are not the natural outgrowth of choice, but that they were dictated by need. It is when dedicated plants do not run to sufficient capacity, because sales were lower than expected, or when the life cycle of a chemical reaches an end and smaller quantities are produced, that the operator is faced with the question of whether the plant can also be used for other products. In this chapter, an attempt is made to present the numerous criteria relevant to these considerations.

I thank everyone who has contributed to this book being accomplished. I thank Prof. Dr. Frey for encouraging me to address the theme of multiproduct plants. My thanks are especially due to Prof. Dr. Wintermantel, who provided the stimulus for this book.

Ludwigshafen, Autumn 2000   J. RAUCH

Editor and Authors

Editor

Dipl.-Ing. JOACHIM RAUCH
BASF Aktiengesellschaft
ZET/ZT – A15
D-67056 Ludwigshafen
Germany

Authors

Sections 3.2, 3.6, 11.1, 11.3
Dr. JÜRGEN CIPRIAN
BASF Aktiengesellschaft
ZET/FA – L540
D-67056 Ludwigshafen
Germany

Chapter 9
Dipl.-Ing. HEINZ-FRIEDRICH ENSEN
BASF Aktiengesellschaft
ZET/ST – L540
D-67056 Ludwigshafen
Germany

Chapter 1, Sections 2.2, 3.1, 3.3, 3.7.5,
4.1 to 4.3, 5.2.2, 5.3, 5.5, 11.2
Dipl.-Ing. STEFAN FÜRER
BASF Aktiengesellschaft
ZET/ZT – A15
D-67056 Ludwigshafen
Germany

Chapter 7
Dr. KLAUS DE HAAS
BASF Aktiengesellschaft
DWL/CB – C10
D-67056 Ludwigshafen
Germany

Chapter 8
Dr. ANDREAS HELGET
BASF Aktiengesellschaft
DWL/LM – L440
D-67056 Ludwigshafen
Germany

Chapter 8
Dr. FRANZ-JOSEF KERTING
BASF Aktiengesellschaft
DWL/LM – L440
D-67056 Ludwigshafen
Germany

Chapter 6
Dr. JÜRGEN KORKHAUS
BASF Aktiengesellschaft
ZEW/BK – L443
D-67056 Ludwigshafen
Germany

Chapter 9
Dr. JÖRG KRAMES
BASF Aktiengesellschaft
ZET/ST – L544
D-67056 Ludwigshafen
Germany

Chapter 1, Sections 3.5, 11.1, 11.1
Dipl.-Ing. JOACHIM RAUCH
BASF Aktiengesellschaft
ZET/ZT –A15
D-67056 Ludwigshafen
Germany

Chapter 1, Sections 2.1, 2.3, 3.1, 3.4, 3.7.1 to 3.7.4, 4.4, 4.5, 5.1, 5.2.1, 5.4, 5.6, 11.2
Dipl.-Ing. FRANZ JOSEF SANDEN
BASF Aktiengesellschaft
DWX/AK – N400
D-67056 Ludwigshafen
Germany

Chapter 7
Dr. JÜRGEN SCHLÖSSER
BASF Aktiengesellschaft
DWL/CB – C10
D-67056 Ludwigshafen
Germany

Section 3.2
Dr. RÜDIGER WELKER
BASF Aktiengesellschaft
ZAV/A – M300
D-67056 Ludwigshafen
Germany

Chapter 7
Dr. DIRK WILSDORF
BASF Aktiengesellschaft
DWL/LP – L426
D-67056 Ludwigshafen
Germany

Chapter 10
Dr. THOMAS WOLFF
BASF Aktiengesellschaft
DUS/AB – M940
D-67056 Ludwigshafen
Germany

Part 1
Basic Concepts

1
Definitions of Multiproduct Plants and Flexibility Demands

1.1
Definitions and Concepts

Multiproduct plants are process plants that can, according to market demand, produce various products. Multiproduct plants are used in the chemical, pharmaceutical, and related industries.

Several different types of plants are considered to be *multiproduct plants*:

In the literature [1.1, 1.2], the concept *multiproduct plant* is used, on the one hand, for a plant that uses a very similar process for the sequential production of different, but generally similar products, belonging, for example, to the same product family. These plants are designed and built, in the first place, for a limited number of products. In reference [1.3], for example, the designation *multiproduct plant* is used exclusively, and more narrowly than we prefer, for this type of plant. The basic type of plant represented in Fig. 1.1 is according to reference [1.2] a *single-line plant*. The small boxes labeled a–f in Fig. 1.1 represent separate process steps or equipment (groups), in which certain unit operations take place.

A second definition of a multiproduct plant (multiline/multipath/network type in Fig. 1.1) includes plants that can produce various products by different processes. This occurs, typically, sequentially, but can also run parallel, where different sections of a plant separately produce different products. For this, the multiproduct plant is divided into independent plant sections. This type of plant is also known as a *multiline/multipath* plant [1.4]. The structure of such a plant is oriented according to the unit operations that are carried out in it. Here the products to be produced may often place completely different demands on the equipment and machinery used. In the literature, this type of plant is often termed a *multipurpose plant* [1.1–1.3, 1.5, 1.6]. Typical for this situation is that the operation units may have different uses and may be circuited according to the technical demands of the process for the product to be manufactured.

In the U.S. literature, the very fitting designation *multipurpose pilot plant* is used for a special type of multiproduct plant [1.7]; the term *general-purpose pilot plant* is also used [1.8]. These refer to multiproduct plants that operate on pilot scale, and are used for process development as well as for producing samples and new products in the quantities required for their introduction into the market.

Single-line

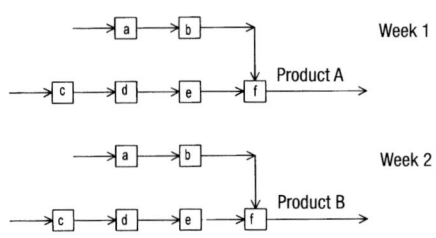

Week 1

Product A

Week 2

Product B

Multiline / multipath / network

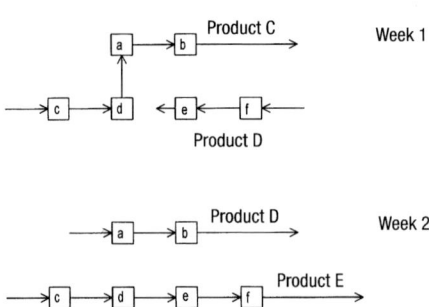

Product C

Week 1

Product D

Product D

Week 2

Product E

Fig. 1.1 Basic types of multiproduct plants, as outlined in reference [1.2] (reproduced with the kind permission of Butterworth Publishers)

It is true that large, continuously operated single-line plants, for example, steam crackers, may also produce more than one product. In such a case, the product mix that leaves the plant may be determined partly by the market (the buyer). In such a plant, one process is used to produce various linked products simultaneously. Such plants are, however, not multiproduct plants according to our definition, since the same product spectrum, even when in a different composition, is produced in such a plant. A discontinuously operated plant, in which the same product spectrum is produced in every charge, is also not a multiproduct plant. These plants are monoplants.

The often used designation *multipurpose plant* is not used in this book, as all the plants considered here serve only one purpose, that of producing products. Only pilot plants serve an additional purpose, that of process development.

A term that has been in recent use is that of *multisubstance plant*. This term appears to originate from the substance-based concepts embodied in the German law on the control of air pollution (BImSchG). The expression *multiproduct plant* has, in contrast, found its way into the NAMUR recommendation NE33 [1.4].

Changing the production with regard to the type and quantity of product produced should not be associated with any or much additional effort or expenditure. Because of the small capacities required and the high flexibility demands, multiproduct plants are often batch plants [1.9]. Even multiproduct plants that operate mainly continuously often have large, discontinuously operated parts.

1.2
Flexibility Demands of Multiproduct Plants

An important characteristic of a multiproduct plant is its flexibility in being adapted to changing requirements. This is because multiproduct plants are employed when the quantity in which the desired product is required is too small to make its production in a monoplant economically feasible. Multiproduct plants are also useful for producing products whose variable market makes the running of a monoplant unfeasible and for the production of several products of similar type (belonging to the same product "family").

In none of these cases is it possible for the plant designer to produce a tailor-made plant, that is, to build a plant that exactly fulfills the requirements for the manufacture of a precisely determined product in a predetermined quantity by a process optimal for this product.

When the products to be manufactured and the processes that are to be carried out in the plant are known, it simplifies matters. It does, however, remain unclear which product is to be produced in which quantities and in what time frame, and if it is to be produced together with other products. As with monoplants, multiproduct plants can, however, be product-oriented or – from the point of the view of the process engineer – be process-oriented. A multiproduct plant that produces products belonging to the same family is such a case, corresponding to the single-line case shown in Fig. 1.1.

It is more difficult to plan and operate a plant that should also produce products not known at the building stage. In such a case, not all the requirements placed by the technology of the process and the properties of the substances are defined. Only the equipment that is required can be considered in the planning of such a plant (equipment-oriented planning). This means that a plant is constructed that contains the equipment suitable for certain unit operations; such a plant can then be used within a wide range of process parameters. This corresponds to the multiline/multipath/network type of plant in Fig. 1.1.

In a publication by Gruhn and Fichtner [1.10], different flexibility types are identified; of these, three types of flexibility are especially important in multiproduct plants:

- *Structural flexibility*. This is the type of flexibility that allows a system, through changes in the connections between its elements, to adapt to changed demands in function. In this way, different processes can be carried out if the connections between the different pieces of equipment in which specific basic operations are carried out are changed.
- *Product-assortment flexibility*. This is the ability of a system to produce different products without the system needing to be changed substantially. This type of flexibility is especially important for multiproduct plants that are used for the production of families of products.
- *Flexibility in capacity*. Such a system can accommodate different capacity demands.

These flexibility types are represented in Fig. 1.2 [1.11].

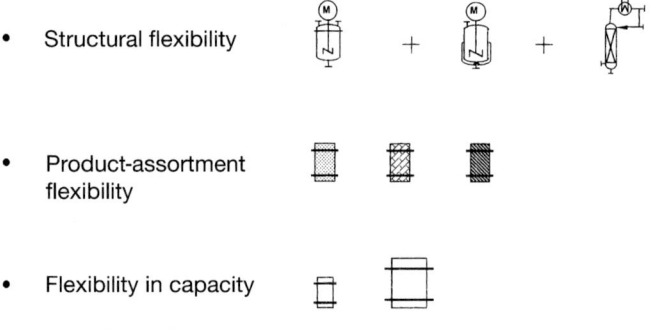

- Structural flexibility

- Product-assortment flexibility

- Flexibility in capacity

Fig. 1.2 Different flexibility types in process plants

Various technical concepts have been developed in the past to meet these flexibility requirements in multiproduct plants.

The flexibility concepts introduced here will be developed further in Chapter 3 (Concepts).

1.3
References

[1.1] PATEL, A. N., MAH, R. S. H., KARIMI, I. A. *Comput. Chem. Eng.* **1991**, *15*, 451–469.

[1.2] MAH, R. S. H., *Chemical Process Structures and Information Flows*, Butterworths, Boston, **1990**, p. 468.

[1.3] SCHUCH, G., KÖNIG, J. *Chem.-Ing.-Tech.* **1992**, *64*, 587–593.

[1.4] NAMUR (Interessengemeinschaft Prozessleittechnik der chemischen und pharmazeutischen Industrie, "User Association of Process Control Technology in Chemical and Pharmaceutical Industries"). Worksheet NE33: Requirements to be met by systems for recipe-based operations. English translation, 19 May 1992.

[1.5] JÄNICKE, W. *Chem.-Ing.-Tech.* **1992**, *64*, 368–370.

[1.6] BERDELLE-HILGE, P. *cav* July **1994**, 51–52.

[1.7] LOWENSTEIN, J. G. *Chem. Eng.* **1985**, *9*, 62–76.

[1.8] PALLUZI, R. P. *Chem. Eng. Prog.* **1991**, 21–26.

[1.9] RIPPIN, D. W. T. *Chem. Eng.* **1991**, 100–107.

[1.10] GRUHN, G., FICHTNER, G. *Chem. Tech.* **1988**, *40*, 505–511.

[1.11] FÜRER, S., RAUCH, J., SANDEN, F. *Chem.-Ing.-Tech.* **1996**, *68*, 375–381.

2
Application Areas

2.1
General

Multiproduct plants are found everywhere in the chemical industry where products in small market quantities but high added values are produced. The yearly production of such products is roughly a few hundred tons, and such products generally have a value of 5–500 Euros/kg. In contrast, bulk products produced in the chemical industry in continuously operating monoplants have market shares of millions of tons per year and their price is typically in the range of 0.5–2.5 Euros/kg.

The typical application areas of multiproduct plants are therefore in the production of fine chemicals, specialty chemicals, and active substances, as well as pharmaceutical agents. Apart from that, such plants are also used to manufacture products according to special requirements for a client. This is, for example, the case with special dyes, plastic blends, or dispersions [2.1–2.3].

Multiproduct plants enable the supplier of the products described above to react quickly to market demands and to fulfill clients' wishes "just-in-time." A multiproduct plant may be comparatively larger and be working to fuller capacity than the corresponding monoplant for an individual product. Someone who runs a multiproduct plant therefore profits from the cost reduction resulting from this.

Small-scale multiproduct plants are also suitable for process development and validation, as well as for the production of samples of new products [2.4, 2.5].

2.2
Research and Development

Multiproduct plants are widely used as pilot plants for research and development. They serve the purpose of process development, dealing with scaling-up questions, and producing samples. Such a pilot plant is referred to in the U.S. literature as a "multipurpose pilot plant" [2.4] or a "general-purpose pilot plant" [2.6]. The pilot scale at which these plants produce is greater than laboratory scale and smaller than that of production plants. Multiproduct plants serve as one of the tools paving the way from the product conception to the construction of the production plant (see Fig. 2.1).

The first phase after the conception of the product idea is that of process screening, where alternative synthetic routes and processes are tested, within the confines set by the requirements and objectives of the project. Laboratory methods and microunits are the tools utilized during these orientational experiments.

In the process development phase, during which the chosen concept for the process is checked, tests are carried out in integrated miniplants [2.7] or in research and development (R&D) multiproduct plants. Small quantities of product already need to be produced during process development, for example, for toxicological or clinical testing in the case of active substances. Miniplants and multiproduct plants are standard parts of research pilot plants.

In the third phase, that of the design of the production plant, the final production process is optimized. Individual equipment that may be scaled up or multiproduct plants are used for working out the dimensions of the equipment and machinery for production.

Once a production process has been decided upon, the product's rapid introduction into the market is of prime economic importance. Multiproduct research pilot plants can play an important role here too.

To keep up with increasing international competition, one needs to keep the time between product idea realization and market introduction as short as possible. This is done by so-called simultaneous engineering, by which the above-described phases of process development overlap as much as possible.

Multiproduct plants used in research need to be versatile, to be able to continuously adapt to the different processes associated with the production of products with different chemical and physical properties. Whereas the development and construction of multiproduct plants used in production are often product-oriented, with known product families produced by similar processes in mind, the planning of multiproduct plants for research need to be technology- and equipment-oriented. This means that the equipment and machinery need to cover certain ba-

Product and process development

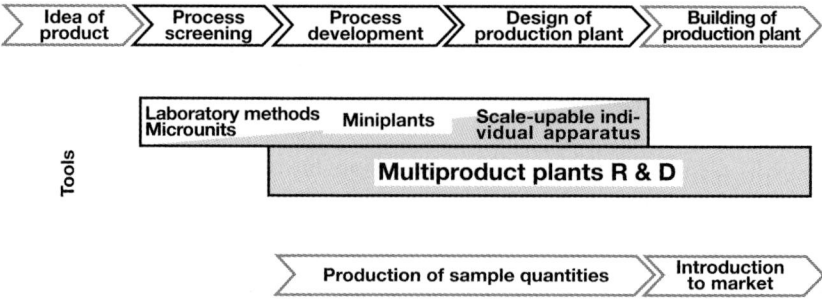

Fig. 2.1 Phases in the development of products and processes

sic operations, with approximately one operation unit for each unit operation. For this, standard equipment, such as agitated reactor vessels, distillation columns, receivers, and so forth, for a broad range of applications (temperature, pressure, capacity, flow rate, corrosion resistance), are used. Enough variation and combination possibilities of the equipment lead to the required structural flexibility for running different processes.

Another difference between multiproduct plants in production and in research is in the extensive instrumentation and process control equipment found in the latter; process development and scale-up of equipment and machinery generally require extra data (see also Chapter 7).

The opposite is found with automation. Because the processes and process parameters that are used in multiproduct plants in production are known, these plants are largely automated, to run as economically as possible. With the constantly changing and still developing processes found in research plants, this is not possible.

2.3
Production

In the following section a few typical areas of chemical production, where multiproduct plants have already found significant application or could find application, will be looked at in more detail. The examples used are not intended to be complete; the intention is merely to illustrate the use of multiproduct plants in production.

Production of fine chemicals and specialty chemicals is one of the most important areas in which multiproduct plants are utilized. The group of fine chemicals is made up of several highly refined products that are either biologically active themselves or are starting materials for such compounds. Fine chemicals are sold according to their specifications and are usually sold by several producers. Specialty chemicals, on the other hand, are sold according to their function, and usually have only one provider [2.8]. The individual products of these types produced in the largest quantities are found among the agrochemicals, especially chemicals used as pesticides. Apart from that, pharmaceutical fine- and specialty chemicals and the products for food and animal feeds also play an important role. From a chemical point of view, the cyclic compounds, especially heterocycles with complicated structures, make up the most important part.

Several of the products that are produced in small quantities for the market are found in this sector. Only a few products from this sector have a market of more than 10,000 tons per year [2.8]. The products have a relatively high price per kg and the demand, especially in the case of agrochemicals, whose demand is partly seasonally determined, tends to fluctuate.

The synthetic routes are often very complicated and differ from product to product. Very high demands regarding purity and environmental friendliness are placed not only on fine and specialty chemicals themselves, but also on their synthetic procedures. This requires not only the highest purity, it also means that cross-contamination between successively produced products is to be avoided. For the protection of those working at such plants and for the protection of the environment, these plants need to meet very high standards, so that the substances, often biologically active, produced therein are handled safely during the production process. This includes the secure containment of these chemicals in the equipment used (see also Chapter 4).

Because of the multitude of possible methods of synthesis, only plants that are characterized by high flexibility in both structure and product assortment are suitable for the production of fine and specialty chemicals. The plants in this area are usually equipment-oriented (see also Chapter 1). This is even more pronounced in the production of pharmaceuticals. Nowadays, it is no longer only the pharmaceutical companies that run multiproduct plants for the production of pharmaceutical fine and specialty chemicals. While the pharmaceutical companies concentrate on their core competencies of discovering and developing active substances, and the galenic formulation of these, they make more and more use of custom manufacturers for the actual synthesis of the active substances [2.9].

Discontinuously operated standard multiproduct plants connected by pipeline manifolds to seldomly used special equipment are suitable for the production of fine and specialty chemicals. Mobile units, such as pumps and equipment for the treatment of exhaust gas or for the dosing of solids may also be used here (see also Chapter 3).

After the actual synthesis of the active substance, a further part of the plant is dedicated only to the mixing and formulation of the active substances and auxiliaries or auxiliary substances. In the pharmaceutical industry this is the galenic domain. Since the active substance is hardly ever marketed in its pure state, the active agent needs to be combined with auxiliary substances and fillers to yield the usable product. The part of the plant that fulfills this function contains equipment for the handling of solids, such as driers, mills, sieves, and mixers, as well as equipment used for agglomerates, such as coating drums or tablet presses, and, finally, also filling and packing equipment. The formulation units are often constructed as multiproduct plants.

A special requirement placed on multiproduct plants in the case of fine and specialty chemicals is the need to be easily cleaned. In the pharmaceutical industry, this requirement stems directly from the GMP (good manufacturing practice) guidelines (of the FDA) that require the prevention of cross-contamination. A technical solution for this problem is that of "cleaning in place" or "cleaning in process" (CIP). This means that the equipment should be cleaned without being dismantled ("in place") or as part of the process ("in process"). For a plant where microbial contamination is an issue, sterilization should be possible. In such a plant, "sterilization in place/process" (SIP) is the equivalent of CIP (see also Chapter 4).

Among paints, dyes, and pigments there are also often large spectra of products (product families) that can be produced according to more or less the same process. That is, certain types of reactions lead to specific product families. In so far, the history of the German chemical industry, as a history of the dye industry, may also be regarded as a history of multiproduct plants. Anthraquinone dyes, azo dyes, and other dye classes are all substance classes that are produced batchwise, according to the same process. This is still valid today in full-scale production.

A good example of this is one of the most important dye classes, that of the azo dyes. Because of the simple synthesis, which normally takes place in aqueous medium, and the practically unlimited choice of starting materials, a broad spectrum of azo dyes is possible; the azo dyes make up the dye class with the most representatives [2.10]. From a primary aromatic amine, in a (hydrochloric) acidic (sodium) nitrite solution, at ca. 0 °C, the diazonium ion is produced. In production, this takes place in an agitated vessel. The coupling components are mixed together in a second agitated vessel, usually together with water and base. The actual coupling reaction only takes place in the third, the coupling vessel (see Figs. 2.2 and 2.3).

Filtration equipment is found between the vessels and after the reaction vessel. This dye synthesis plant corresponds to the concept of the batchwise operated standard multiproduct plant or an adapted version of the standard multiproduct plant (see also Chapters 1 and 3).

Industrial complexes that produce azo dyes usually contain several lanes of vessels in various sizes ($25\,m^2$ to $80\,m^2$). Individual lanes are, in part, also used exclusively for the production of specific color tones. This avoids intensive cleaning operations and the danger of cross-contamination.

Other reaction vessels could also follow on from the coupling vessel, to carry out further reactions such as saponification, esterification, or acylation.

If the pure pigment is not yet a marketed product, the formulation stage then follows, as in the case of the production of active substances. During this stage, the pigments and various auxiliaries are ground, mixed, dissolved, and so forth, to formulate products that can be sold.

The formulation of a dye is schematically represented in Fig. 2.4 [2.11].

After weighing and dosing of the individual components, the crucial step in the production of dyes is the complete wetting and dispersion of the pigments. For the pigments to be well dispersed, an agitated vessel is not sufficient, and rotary wet-grinding mills and ball mills are therefore used. This mill can be operated within the recycle phase of a stirred vat or, more effectively, used in single-pass between the disperser and mixer vessels.

The sequence of operations followed in such a plant depends on the product. If only intermediates are produced, the process ends after dispersion, and the disper-

$$Ar-NH_2 + XNO + RH \xrightarrow{-H_2O} Ar-N=N-R + HX$$

Ar: Aryl group
X: Cl, Br, NO_2, HSO_4

Fig. 2.2 Reaction scheme of the production of an azo dye

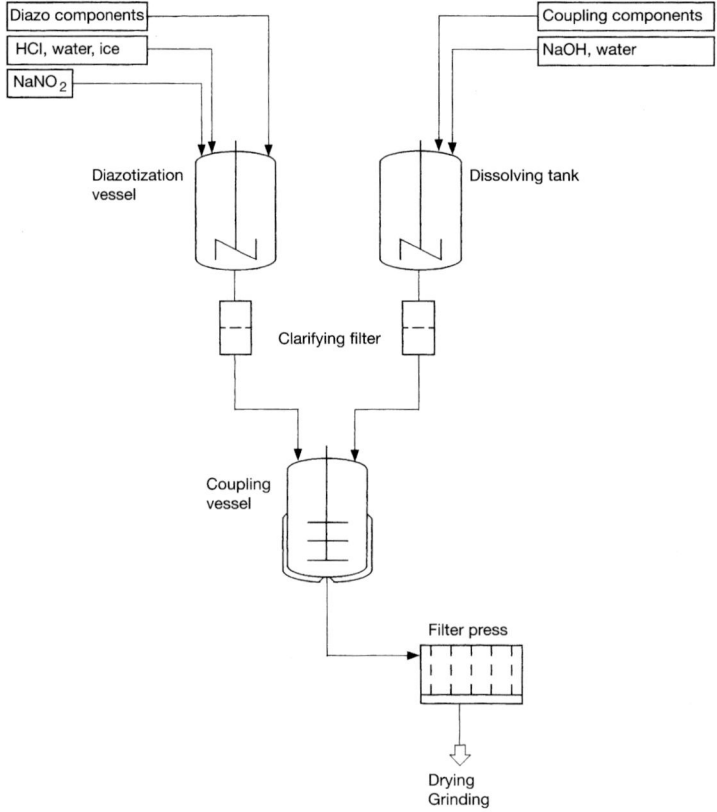

Fig. 2.3 Schematic representation of a plant for the production of azo dyes, according to reference [2.10]

sion thus produced is sent for storage or is packed. To produce a dye for the market, further solvents, binders, auxiliaries, and monomers still need to be added. Before the final packing of the marketable product, there is often a final filtration step. Due to the great number of feedstocks, fillers, and final products, high logistical flexibility demands are placed on such a plant. In comparison, the technical demands placed by the processes are not very high, as all products can be produced by a limited number of unit operations. The plants in this area are usually set up to be product- or process-oriented.

The production of lubrication oils or fats is characterized by an immense spectrum of products for a countless variety of applications. Mineral oils, the distillation products of petroleum, can be improved by further steps, such as hydration, oleum refinement, and the removal of tars and waxes. The properties required by the desired application can be obtained through mixing of various oils. For the mixing of various crude oils, synthetic oils, and additives, pipeless plants are used (especially in Japan). The only technical process carried out in such a plant is that

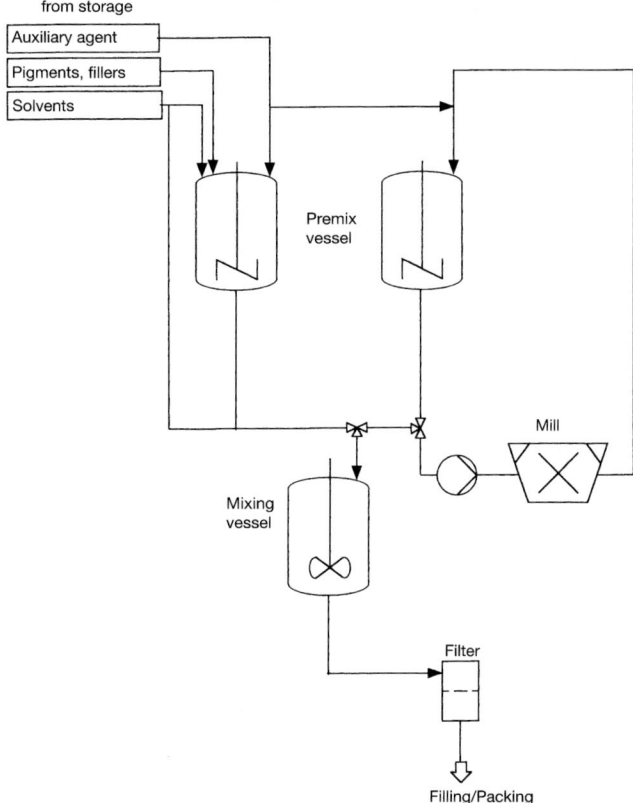

from storage

Auxiliary agent

Pigments, fillers

Solvents

Premix
vessel

Mill

Mixing
vessel

Filter

Filling/Packing

Fig. 2.4 Schematic representation of the formulation of a dye, according to reference [2.11]

of mixing. The components are charged into mobile mixing vats at the filling station, mixed at the mixing station, with heating or cooling where necessary, and then passed on to the discharging station for packing. The mobile containers are cleaned thoroughly at the cleaning station, and, with the danger of cross-mixing thus avoided, the cycle can start anew (see Chapter 3).

A great part of the production of lubrication fats consists of the manufacture of soaps and the mixing of these with oils. Their production mostly takes place in discontinuously operating multiproduct plants. Fig. 2.5 shows, based on reference [2.12], the principles on which a lubrication fat plant is constructed. Such a plant is centered around a special agitated vessel with a heating mantle and built-in heat exchanger, as well as a connected agitated vessel with oppositely operating double stirrers and auxiliary equipment. In the first step of the process, the fatty acids, mixed in a part of the oils, are charged into the first reactor. After addition of the hydroxides, dissolved or suspended in water, the conversion into soap takes place. In the second container, the rest of the oils and additives are added, water

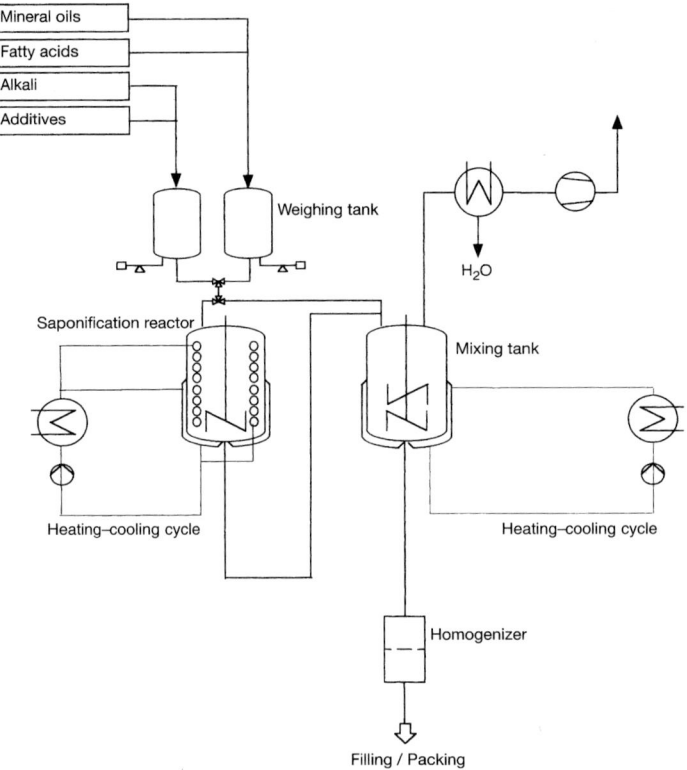

Fig. 2.5 The basic construction of a lubrication-fat plant, according to reference [2.12]

is removed, the soap is crystallized, and the product is homogenized. For various product lines, it may make sense to use different vessels for this second step. Depending on what additives are added (graphite or molybdenum compounds), the fats obtained may be "black" or "white." According to our classification system, this plant is a discontinuously operated standard multiproduct plant; its operation requires only minimal process flexibility (see also Chapter 3). The demands on its logistical flexibility grow as the product variety increases. This is also a case where the plant is set up to be product- or process-oriented.

2.4
References

[2.1] JÄNICKE, W. *Chem.-Ing.-Tech.* **1992**, *64*, 368–370.

[2.2] NIWA, T. *Chem. Eng.* **1993**, *6*, 102–108.

[2.3] JÄNICKE, W., SCHULZE, J. *Chem.-Ing.-Tech.* **1993**, *2*, 193–195.

[2.4] LOWENSTEIN, J.G. *Chem. Eng.* **1995**, *23*, 62–76.

[2.5] PALLUZI, R.P. *Chem. Eng. Prog.* **1991**, *1*, 21–26.

[2.6] PALLUZI, R.P. *Chem. Eng.* **1990**, *97(3)*, 76–88.

[2.7] BUSCHULTE, T.K., HEIMANN, F. *Chem.-Ing.-Tech.* **1995**, *67(6)*, 718–724.

[2.8] POLLAK, P., in *Kirk–Othmer Encyclopedia of Chemical Technology, Vol. 10, 4th ed.*, Wiley, New York, **1994**.

[2.9] SCHREINER, G., WIDMER, A. *Chem. Rundschau* **1996**, *24*, 2.

[2.10] HUNGER, K., MISCHKE, P., RIEPER, W., *Azo Dyes*, in *Ullmann's Encyclopedia of Industrial Chemistry, Vol. A3, Chap. 1, 5th ed.*, Wiley-VCH, Weinheim, 1985, p. 245.

[2.11] HILLER, R., *Paints and Coatings*, in *Ullmann's Encyclopedia of Industrial Chemistry, Vol. A18, Chap. 7, 5th ed.*, Wiley-VCH, Weinheim, **1991**, p 474.

[2.12] KLAMANN, D. *Lubricants and Related Products*, in *Ullmann's Encyclopedia of Industrial Chemistry, Vol. A15, 5th ed.*, Wiley-VCH, Weinheim, **1990**, p. 423.

3
Concepts

3.1
The Discontinuously Operated Standard Multiproduct Plant

3.1.1
General

The standard discontinuously operated multiproduct plant is very similar to a conventional batch plant, the most frequently encountered and oldest type of multiproduct plant, of which there are many variations. The conventional batch plant is product-specific, and contains only the equipment necessary for the production of a specific product; in contrast, the likelihood that a standard multiproduct plant will be used for many different tasks is incorporated into the planning from the outset. The extensive technical equipment that is available may not always be in use, but is available should it be required. Equipment suited to a wide spectrum of product properties and process parameters is usually preferred. Such plants are particularly useful for custom manufacturers [3.1].

A significant characteristic of this type of plant is the permanent connections between the tubing of the equipment. Cleaning between product changeovers is then a very complex problem. In practice, the whole unit is often cleaned by a suitable solvent being boiled in the plant; this is a very intensive process. A proposed solution to this problem has been that of permanently installed CIP (cleaning in place) installations [3.2].

3.1.2
Structure of the Plant

3.1.2.1 Basic Construction
The center of the plant is an agitated vessel with a distillation column placed on top of it and a secondary heating–cooling cycle [3.3, 3.4]. The structure of such a discontinuously operated standard multiproduct plant is represented in Fig. 3.1. For the handling of feedstocks, products, and solvents, a sufficient number of loading vessels, buffer containers, and receivers are at hand. Each peripherally placed buffer vessel can receive the total content of the reactor. Agitated reactor vessels and peripheral equipment are connected to each other by fixed pipes.

Loading vessels

Agitated reactor vessel
+ columns

Secondary heating–cooling cycle

Inverting filter centrifuge

Drier

Receivers

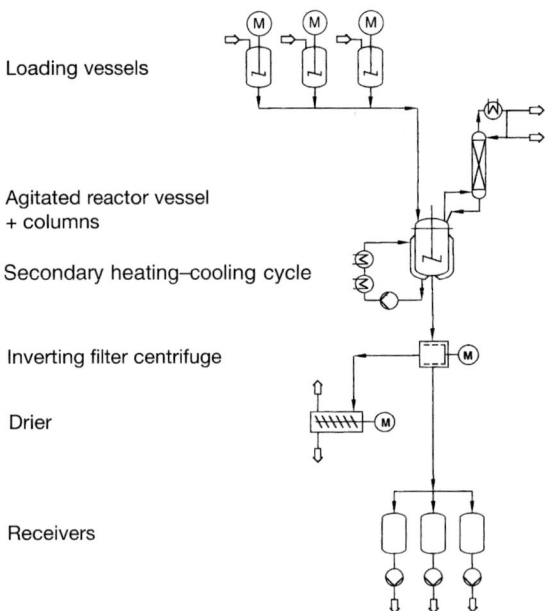

Fig. 3.1 Standard multiproduct plant

The columns on top of the reactor are for simpler distillations, such as removal of solvents or reaction side products. This task is usually fulfilled by glass columns with a few meters of random packing or structured glass or ceramic packing.

The vertical positioning of the apparatus allows a gravitational flow direction of the substances: from the loading vessels and batch vessels into the agitated reactor vessel, placed below them, and then further into the receivers placed underneath the reactor. Despite this vertical arrangement, it is still useful to equip central apparatus such as stirring vessels and important receivers with pumps and pump cycles. This prevents difficulties with highly viscous media or slurries.

Equipment and machinery for solid–liquid separations are usually found below the agitated vessel. Solid–liquid separations are required for the workup of solid-containing reaction mixtures or the separation of catalysts in the case of suspension-catalyzed reactions. At this point, centrifuges such as inverting filter centrifuges or agitated pressure nutsch filters that are adaptable may be incorporated. Below the filter, a drier that is as widely applicable as possible, for example, a shovel drier, is chosen, to ensure independent drying if needed (see Section 3.6).

The pumps that are in use should be corrosion-resistant and be adaptable regarding rate and pressure of delivery (see also Section 4.5).

Product-specific components, which can only be used for the production of a single product, are also sometimes found within multiproduct plants. These are not in use during the manufacture of other products. Examples are a photoreactor incorporated into the recirculation cycle of an agitated vessel, and a mixer–settler setup, operating as a multistage extraction unit, placed below the agitated reactor vessel. Such special equipment, used only partly, is a typical component of a standard multiproduct plant.

3.1.2.2 The Agitated Reactor Vessel as Central Apparatus

The three basic reactor types used in chemical processes are the following:

- Pipe reactors
- Continuously operated agitated reactor vessels
- Discontinuously operated agitated reactor vessels

The central piece of equipment, the heart of the discontinuously operated standard multiproduct plant, is an agitated reactor vessel. This is mainly because of the product-assortment flexibility attainable with such a vessel. The ability to be

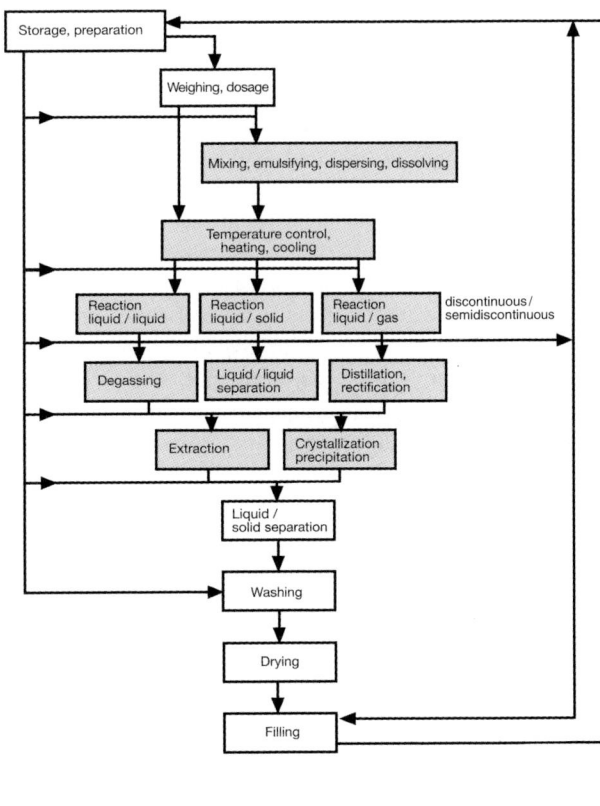

Fig. 3.2 Flow chart of the possible processes in a discontinuously operated multiproduct plant

carried out in a discontinuously or semidiscontinuously operated agitated reactor vessel is characteristic for the reactions that can be carried out in discontinuously operated standard multiproduct plants. Due to the limited heat transfer in discontinuously operated agitated reactor vessels, only reactions that are slow or have low heat of reactions can be carried out in them.

This can be improved by the use of external heat transfer circulation. Another possibility is to run the reaction under reflux. The third possibility is to dilute the reaction mixture with a suitable solvent. Examples of reactions carried out in agitated reactor vessels are the following: polymerization (bulk, solution, and emulsion), hydration, saponification, nitration, acylation, and oxidation. Reactions in which suspended catalysts are exposed to gas are semidiscontinuous, as are reactions to which solids are added. Obviously, most of the unit operations typical for multiproduct plants listed in Fig. 3.2 can be run in agitated vessels.

3.1.3
Application of Discontinuously Operated Multiproduct Plants

The typical size of a central agitated reactor vessel is 4–12 m^3; the diameter of the column placed on top of it is typically 300–500 mm. The sizes of the remaining equipment are based on these central items.

In discontinuously operated standard multiproduct plants, temperatures of –15 to 200 °C are accessible. Heating and cooling often take place through the secondary heating–cooling cycle. The apparatus are often operated under vacuum, at normal pressure, or at slight overpressures (ca. 10 mbar to 3 bar).

The materials used in multiproduct plants should be sufficiently chemically resistant to cope with the great variety of tasks undertaken in these plants. Glass-lined steel, glass, and Hastelloy steel are generally used. Equipment such as tubing and fittings (valves), in particular, that are coated with highly fluorinated materials (PFA and PTFE) are gaining in importance (see also Chapter 4). The materials used in the sealants must be as widely applicable as those found in the rest of the equipment; PTFE- and graphite-based materials are widely used.

The measuring and regulating instruments in standard multiproduct plants also need to be able to fulfill a wide range of possible applications, measurements, and processes. The broad application range of certain chemical engineering equipment and machinery is, however, so far not adequately covered by the available measuring instruments. For example, the throughput of a distillation column that can be operated under vacuum or at normal pressure can vary much more than what can be covered by the measuring range of a flow-rate detector. In such a case, the measuring instrument needs to be exchanged, or its measuring converter should at least be adapted.

The coherent automation of a multiproduct plant requires a unit operation concept (recipe) to be represented in the distributed control system. Such concepts are available [3.5], but are seldom realized (see Chapter 7).

If the plant is set up for a specific product range, the application areas described above can be extended in some directions. For example, in a multiproduct plant for hydration processes, where hydration vessels comprise the central equipment, higher operating pressures could be planned in.

Custom manufacturers usually have several production trains with differing corrosion resistances available in their multiproduct complexes. A train with a glass-lined steel agitated reactor vessel, mounted with a glass column, and equipped with a Hastelloy nutsch filter may be found adjacent to a train constructed of stainless steel.

The fixed connections between the equipment in a standard multiproduct plant lead to low structural flexibility; the same goes for its flexibility with regard to capacity. Its product-assortment flexibility, on the other hand, is high.

In practice, a standard multiproduct plant is a mixture of elements from modular multiproduct plants and pipeline manifolds.

Standard multiproduct plants could surely undergo further developments, owing to the improved possibilities with regard to cleaning between product changeovers. Although contemporary plants usually contain standard chemistry equipment, the first effects of the CIP concept (see Chapter 4) can be seen in, for example, the construction of centrifuges and process filters. Improvements in the design of agitated reactor vessels and receivers, leading, for example, to pocket-free constructions and smoother surfaces, are also foreseeable. The integration of tubing for CIP liquids into the equipment should also be a source of future improvements [3.2].

3.2
Continuously Operated Standard Multiproduct Plants

For an overview of the long-term significance of continuously operated multiproduct plants, please see reference [3.6].

Continuously operated multiproduct plants are characterized by higher production volumes, production taking place within different campaigns, a large variety of products, albeit of only one or a few product classes, and a fixed plant configuration. The extents to which at least the main products of a continuously operated multiproduct plant are produced should be sufficient to justify their production in a continuously operated single-product–single-line plant; the crucial reason for choosing a multiproduct plant is that further (and smaller) product volumes will often need to be produced, by an identical synthetic route and process, but the production of the smaller products on their own would not have been economically feasible. Therefore, the campaign periods are longer for the main products and shorter for the minor products. Although such a plant has a fixed layout, different synthetic and/or workup routes should be possible, so that, depending on the product requirements, all parts of the plant are not always in use.

Developing a new continuously operated multiproduct plant is not so much the creation of a new plant as it is the optimization of an already running continuously operated multiproduct plant and the integration of the production of new products into such an existing plant. The approach to this [3.7, 3.8] is generally

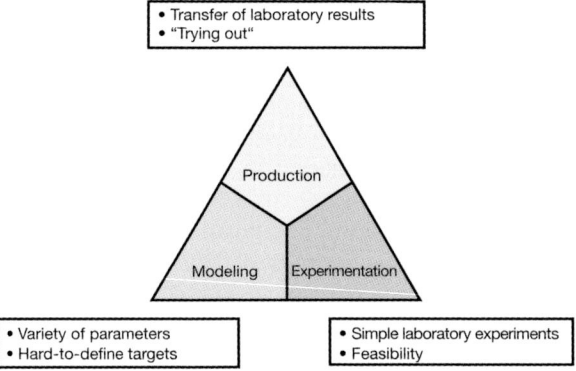

Fig. 3.3   The utilization of resources in multiproduct plants

very similar to that taken with continuously operated single-line plants, and is briefly dealt with in Chapter 11. No more will be added here except to say that a plant that is generously equipped and has a good assortment of apparatus as well as the possibility of being extended by further identical production lanes is the only one capable of realistically meeting the demands of high flexibility regarding product variety and production capacity.

The economically feasible integration of the production of new products into an existing continuously operated multiproduct plant is only possible if the process information is available and interpreted; ideally this information base should be complimented by simulation of the process after adjustments have been made for the task at hand, or by experiments carried out in a laboratory or even in the plant. This iterative procedure of process development combined with modeling, experiments, and production (Fig. 3.3) extends the knowledge about the process; modeling and screening experiments help towards understanding the relevant issues and to develop new approaches to improved processes, which can again be applied directly to the operation. In this way, new products can reach the market more rapidly. This influence can extend from the point where the need for a new product is recognized, via the creation of the operation guidelines, the production of sample quantities, up to the actual production point [3.9, 3.10].

This approach is also typical for developing a new idea for a continuously operated multiproduct plant. Normally, when a new multiproduct concept has already existed in a similar form, or if individual products have already been produced in other plants, the relevant data and information are already available long before the phase of developing a new concept has started.

3.2.1
Plant Structure

The individual products – taking up differing extents of the production and facilities – are produced in suitable, and often fixed production lines. Discontinuous processes may be combined with continuous ones, and the feedstocks, workup of mutually useful substances, as well as waste-gas treatment and disposal of effluents and residues are usually at common disposal.

For the manufacture of individual products, continuously operated multiproduct plants are at a clear disadvantage compared to single-product, single-line, and discontinuously operating plants: the production of the individual products cannot be optimized as in the case of single-product plants, since the requirements of an entire class of products need to be met with first. Moreover, only a pool of equipment with limited flexibility is usually available for coping with the requirements for producing specific products.

Fig. 3.4 shows the structure of the most complex case of a continuously operated multiproduct plant. Each functional unit – synthesis, waste-gas condensation, scrubbing, phase separation, and workup – could be characterized by various types of equipment. The synthesis unit could, for example, consist of discontinuously and several continuously operated reactors; the production lanes may contain distillation columns for workup that can be optimized for the various product groups with regard to separation ability, throughput, product discharge, and links. The individual products – with different properties and requirements – could follow different routes through a continuously operated multiproduct plant.

To realize an accordingly high structural and product-assortment flexibility under such complex circumstances, a greater number of individual pieces of apparatus is required, to extend the number of potential connection and operating possibilities. The limited flexibility of the individual apparatus can be compensated for

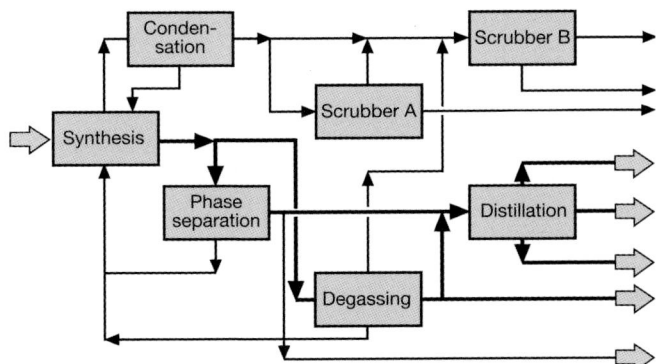

Fig. 3.4 Schematic representation and synthetic routes of a multiproduct plant

by the introduction of several interaction possibilities. This complexity also places demands on the facilities for the storage and loading of feedstocks and products.

Continuously operated multiproduct plants are often found at integrated production sites, for the sake of cohesion, the availability of feedstocks and auxiliaries, flexibility, the safeguarding of production, and so forth.

3.2.2
Technical Limitations

The output spectrum of continuously operated multiproduct plants can easily span up to 50 individual products. Only a limited number of product classes can, however, be produced in a continuously operated multiproduct plant. The processes used for the production of the different product classes will have significant similarities, for example, identical or comparable synthetic routes, identical feedstocks, same separation characteristics, and the physical material data would also be within similar limits.

Liquid material systems dominate with regard to equipment technology and product changeover. With the technology available for solid operations, the individual basic operations are invariably more expensive than those for fluid operations. Moreover, with the equipment used for solid operations, bottlenecks are soon encountered if the product palette is changed, as the application range is limited because of higher specialization. In principle, however, the different states of aggregation encountered in continuously operated multiproduct plants can cover the whole spectrum from liquid to solid. This is not only the case for the main products, it is also valid for feedstocks, side components, and residues.

3.2.3
Plant Types

Continuously operated multiproduct plants can be divided into three basic types, which differ not only with regard to equipment and products, but also with regard to their structural and product-assortment flexibilities.

3.2.3.1 Continuously Operated Single-Line Multiproduct Plants for a Small Number of Very Similar Products (*Type 1, Synthesis-Oriented*)

Type 1 continuously operated multiproduct plants, usually designed for high capacities, are similar to single-product plants, but are not product-optimized. Operating parameters such as pressure, temperature, or recycling can be changed in this type of multiproduct plant, to adapt it to the product required. The connections between the apparatus are, on the other hand, fixed, as whole functional units are usually required. The technical unit operations are mainly determined by the product synthesis undertaken and the additional processes that follow from that. Typical examples of processes carried out in these multiproduct plants are chlorination, nitration, or hydrogenation reactions [3.11], as well as certain alcohol distil-

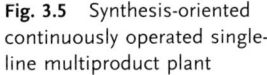

Fig. 3.5 Synthesis-oriented continuously operated single-line multiproduct plant

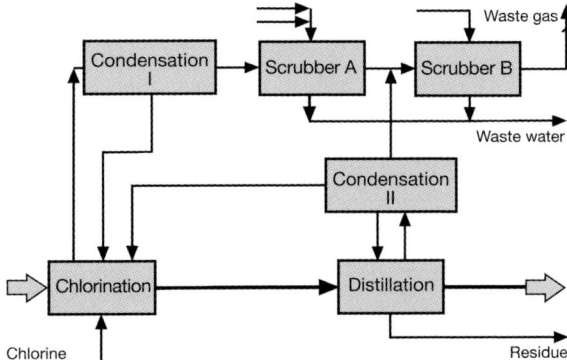

lations, where the product class is the deciding factor. An important characteristic is that the technology and equipment used are always directed at the pertinent synthetic method or corresponding unit operation (Fig. 3.5).

3.2.3.2 Continuously Operated Single-Line Multiproduct Plants for a Single Product Class (*Type 2, Product-Class-Oriented*)

Type 2 continuously operated multiproduct plants are single-line plants whose apparatus are linked in a fixed, unchangeable configuration. They are used for the production of a large number of products belonging to the same product class, and therefore producible by the same synthetic route and workup procedures. The products will also, within relatively narrow limits, have comparable material properties, have similar specifications, and identical application areas. The operation units are mainly determined by the requirements of the product class, whose individual members need to follow a defined sequence of further treatments if the desired properties are to be obtained. Typical examples of procedures suitable for this type of multiproduct plant are aminations or similar reactions, and the preparation of certain polymers or acid chlorides.

An important characteristic of such a plant is that the technology used as well as the product class, that is, the application area of the products, are always the same. This means that the process engineering structure of this plant type can be the same as that of the previously described type; the crucial difference lies in the purpose of the products. What distinguishes this plant type is that its overall concept is not only determined by the synthetic route, but also by identical workup and purification steps, because the products belong to the same class. This type of continuously operated single-line multiproduct plant is the one encountered most often (Fig. 3.6).

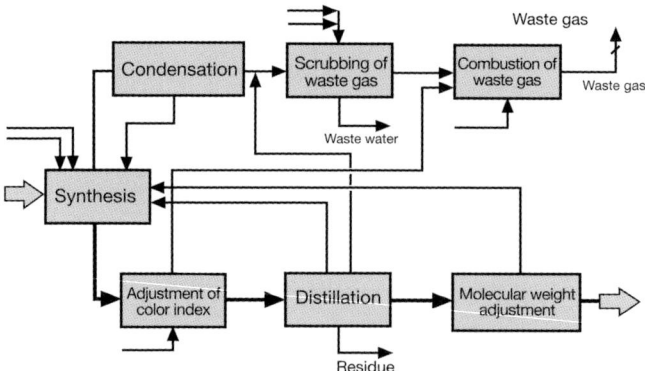

Fig. 3.6 Product-class-oriented continuously operated single-line multiproduct plant

3.2.3.3 **Continuously Operated Multiline Multiproduct Plants for More Than One Product Class (*Type 3, Synthesis- and Product-Class-Oriented*)**

Type 3 continuously operated multiproduct plants are very complex with regard to apparatus connections and product variety, and are characterized by several limitations and structural conditions. The basic structure is that of a continuously operated single-product plant (cf. Fig. 3.4), but with one defining difference: not every available unit operation or equipment unit is always used (Figs. 3.7 and 3.8).

The equipment are characteristically combined into functional units, through which the different products proceed in different ways. The various individual representatives of a product palette may undergo the same basic operations in different functional units, which are accessed through different linkage configurations. Some entire functional units, such as those for scrubbing, degassing, or distillation may not be necessary at all for some products. Individual pieces of apparatus may also provide product-specific routes, for example, the product outlet

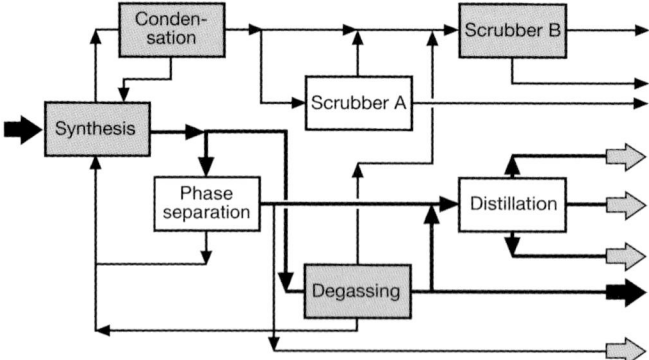

Fig. 3.7 Synthetic route for product 1 in a continuously operated multiline multiproduct plant

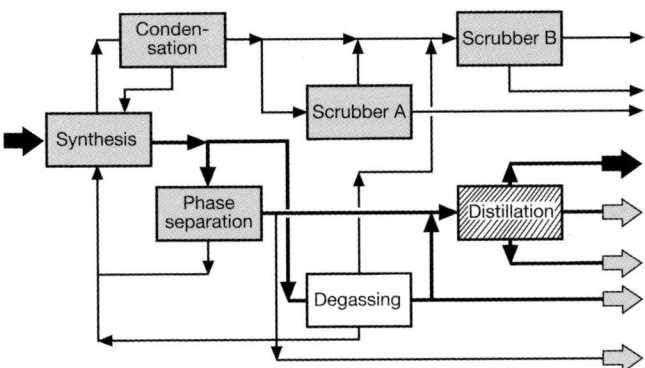

Fig. 3.8 Synthetic route for product 15 in a continuously operated multiline multiproduct plant

may be the top-, bottom-, or side stream of a distillation column, as determined by the vapor–liquid equilibria with the solvent and other components.

This flexibility is restricted by several factors. Every product is subject to the boundary conditions of the available equipment and all the other products, and also places its own material-based demands, thereby influencing the whole production. Very high demands are therefore placed on the management of the entire product palette, from planning to running the process. This type of plant is primarily characterized by the fact that both the technology and the product class, that is, the application areas of the products, may not be identical, but are relatively similar, and that not all the items of available equipment are always in use.

Some versions of this third type of plant also conform to the definition given in Chapter 1 for a multiline–multipath plant.

3.2.4
Examples of Processes in the Different Plant Types

Examples of continuously operating single-line multiproduct plants for a small number of very similar products (type 1 according to the classification above) are plants for chlorination, nitration, hydrogenation, the production of isomers (where, e.g., one isomer is the starting material for a dye, the other is the starting material for a pesticide), the production of differently substituted products from the same feedstocks, or the synthesis of products that differ in molecular chain lengths. A borderline case is that of coupled production, where two or more products are produced simultaneously, and the product proportions can be varied according to market demand, by variation of the operating conditions.

The production of esters is an example of the use of continuously operating single-line multiproduct plants to yield an entire product class by an identical process and for the same use (type 2 according to the classification above). Involved are acid-catalyzed equilibrium reactions, where the water formed by the reaction is continuously removed to influence the equilibrium, workup is usually by distillation, and recycling of feedstocks is common. Since many esters are used as solvents, for example, for dyes or dispersions, this also means that the specifications are similar. This type of plant is also often used for the production of polymers, for example, PTHF (polytetrahydrofuran).

The more complex type of continuously operated multiline multiproduct plant (type 3 according to the classification above) is found in the production of numerous product or reaction classes, such as amines, amides, vinyl ethers, acrylic esters, or plasticizers.

3.2.5
Example of a Process Modification

The following example, showing typical marginal changes in the production conditions, is used to show the typical characteristics of the latter type of multiproduct plant [3.9, 3.10].

In a continuously operated multiline multiproduct plant, several production lanes (which may also be interconnected) are used to produce approximately 40 products belonging to three product classes. Reactions involving corrosive mixtures and exothermic reactions, with massive gas development, need to be managed for this process. Proper management of thermal balance is crucial: a large part of the heat generated by the reaction is removed through condensation; feedstocks and products lost with the waste gases are highly volatile and therefore need to be condensed at low temperature.

The infrastructure of the process is based on the use of a uniform coolant, which is used in several parts of the plant: during synthesis, workup, and wastegas treatment. This coolant is harmful to the ozone layer, and it is therefore necessary to find a substitute for it.

Different coolants have, however, different thermodynamic properties, and there is no substitute coolant with which the process can be run under the same conditions as before; the conditions will therefore have to be adapted for the use of the new coolant. A suitable substitute for the coolant can only be chosen on the basis of the identification and evaluation of the important operating data, as organized by the following themes:

- Reaction flow
- Thermal balance
- Materials
- Handling of substances
- Product quality
- Product campaign, product changeover
- Availability
- Waste disposal, etc.

From the systematic production and equipment tables, the key products for specific characteristic or critical production properties can be identified in the different equipment units, so that analogies can be drawn.

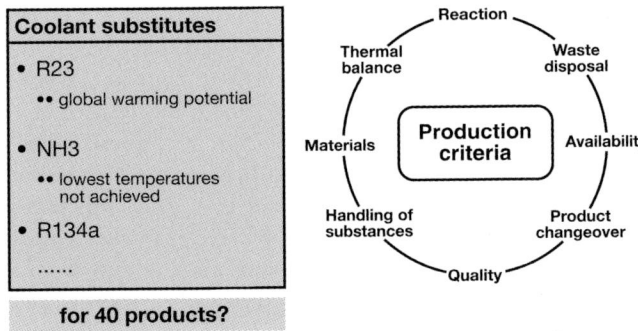

Fig. 3.9 Finding a coolant substitute that best fulfills the production criteria

For further evaluation, it is useful to model the production process of the most important key products with their respective main components, using easily determined or well-known material data and parameters and data from the actual process. From this, a model simulating the multiproduct plant in which the process-relevant steps are carried out can be produced.

Gaps in the description of the process, for example, factors that are only afterwards recognized to be crucial determinants of the reaction, can be filled in by screening or basic experiments or by additional determinations of material properties.

The steps required for the introduction of the substitute coolant, from modified process parameters and adjusted campaign plans, up to changed logistics, can be determined on this basis. The principles of this process are represented in Fig. 3.9.

3.2.6
Special Aspects of Process Engineering

Most of the publications on this subject deal with optimization of the campaign sequence, use of equipment, and similar issues [3.12–3.14]; very little has been published on devising continuously operated multiproduct plants [3.15].

Because of the great variety of products and the sometimes complex operation of the equipment, knowledge about the processes taking place in the plant and the product properties is crucial. There is a lot of information available on the properties of materials in technical equipment, their operation and maintenance, and the handling of products. Most important is knowledge about the critical pathways of a continuously operated multiproduct plant; that means identifying products that are representative of a product group, and products in which the currently used equipment has reached their operating limits. This information should be available to plant operators in a systematic and schematic form

(Fig. 3.10). Only after consequent application of the experience gained by production is it possible to transfer laboratory results, which merely indicate whether something is possible in theory, directly to the plant. The need to go directly from the laboratory to production often results from the lack of opportunities to run experiments, especially for the production of representative sample quantities (in such a case, experience with the particular type of reaction involved is usually, at least, available instead). Other causes may be insufficient information about and around the new products, such as life cycle, market potential, and specifications, or changes in the incidental conditions, such as new laws, or changes in the availability and/or the quality of the feedstocks, energy, and auxiliary substances, which can affect not only the individual product pathways, but also the complete product palette.

The systematically characterized operation process of a continuously operated multiline multiproduct plant forms the basis of an economically viable process management, and may also form the basis for optimizing an existing continuously operated multiproduct plant, as in when a new product needs to be integrated into a preexisting plant.

Abstraction of the operating data of the key products gives the characteristics of an operation, so that:

- Complete data sets for evaluating the process are available
- Limiting equipment and/or products or product pathways are identified

The following should be known: which process steps are involved, what equipment is needed, what are the process conditions, and, especially, what are the effects on the preexisting product palette, and, finally, whether or not these factors necessitate further restructuring as a result of new or preexisting limitations.

Suitable production lanes can be devised in detail by analysis and comparison. The operating conditions, maximal throughput, losses, and operating costs can be determined by simulations. The instructions for operation, product changeover, and cleaning up can be developed on this basis. The same steps are basically re-

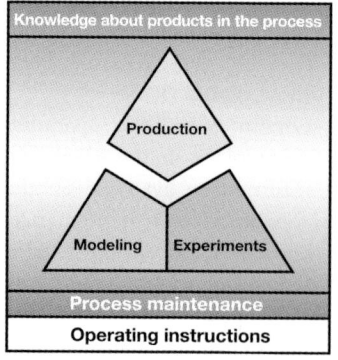

Fig. 3.10 Gathering knowledge for devising a multiproduct plant

quired again if the procedure is modified when the same production lane is used for other products.

The whole production palette needs to be re-examined for the purpose of possible process revision if, after integration of a new product into an existing continuously operated multiline multiproduct plant, checking of the individual lanes, and optimization as described above, no capacity is found for the new product. This may, in some cases, require investments to be made in equipment, if suitable equipment is not part of the basic layout of the plant; this solution at least avoids the need for setting up a completely new plant, and thereby guarantees the necessary flexibility.

The above discussion shows the complexities and difficulties associated with the integration of new products and the optimization of the production of existing product palettes in continuously operated multiline multiproduct plants.

3.3
Modular Multiproduct Plants

3.3.1
Definitions

The terms "module" and "modular plant technology" have been used to describe various technical concepts. In an example from the literature, which deals with the subject of plant structures for high-value products manufactured by procedures needing extensive know-how, a compact plant was called a module [3.16]. Such a "module" is an independent plant section of compact construction that is integrated into the total infrastructure, as, for example, air-separation, absorption, or heating units. In this sense, a "module" is usually not part of a multiproduct plant.

In another example of what is understood by this concept, the designation "modular construction" has been used for the preassembly of whole, not independent parts (process equipment) for conventional chemical plants at suitable assembly locations, including off-site [3.17]. The advantage of this over conventional on-site construction is that time and costs are saved, because preassembly of the transportable plant parts (modules) can take place in an optimal work environment and under optimal conditions. Several plant parts may also be assembled simultaneously. This method is more efficient and safer than exclusively on-site assembly. The labor costs are also generally lower.

In another interpretation of the concept, pilot plants for specific and well-known tasks (see description in Section 2.2), are made up of assembled "modules" and are placed as preassembled units in a single large apparatus rack at the required location [3.18]. In this sense, every "module" represents one or more process functions or unit operations and is widely standardized. Process functions that have been implemented as pilot-plant modules are, for example, gas dosing, gas scrubbing, liquid dosing, loading of solids, preheating/evaporation, overheating, inertizing, reactions, sampling, normal-pressure distillation, vacuum distillation, and gas/liquid separation. The factors determining the choice and combination of suitable modules for different pilot plants are mainly pressure and temperature limits, filling volumes, flow quantities, and the physical properties of the liquids.

In contrast to the rather divergent meanings attached to modular plants, as described above, the concept "module" is used differently in the following definition: *a module is a component of a modular multiproduct plant with which one can very flexibly produce different products by different processes.* Modules are represented by compact, usually mobile, ready-to-use functional units (apparatus, machinery, etc., also in combinations), which cover different unit operations in process technology.

3.3.2
Plant Structure

Modular multiproduct plants often consist of a permanently installed central piece of equipment, an agitated reactor vessel with secondary heating/cooling circulation and mounted with a distillation column with condenser (Fig. 3.11). The required plant structure, meeting the needs of the process, is easily assembled from different modules, grouped around the reactor. In such a case, the modules cover a wide range of process engineering functions (e.g., loading, dosing, discharging, heating/cooling, evacuating, pumping, gas scrubbing, filtering).

The modules and the permanently installed parts (process equipment) are usually connected by flexible tubing and standardized connection systems (see

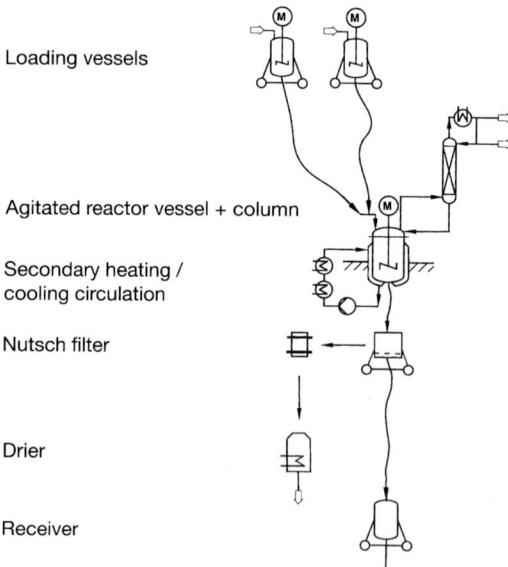

Loading vessels

Agitated reactor vessel + column

Secondary heating /
cooling circulation

Nutsch filter

Drier

Receiver

Fig. 3.11 Structure of a modular multiproduct plant

also Sections 5.2 and 5.3). In this way, the amount of craftsman's input required for adapting or changing over a plant for another product is reduced to a minimum.

Suitable spatial arrangement of apparatus and machinery within a plant allows the product flow to proceed mainly by gravity. A two-storied building facilitates an agitated vessel cascade setup so that, for example, the first agitated reactor vessel is used for the reaction and a phase separation, and the second serves for the subsequent workup stages (e.g., rectification, crystallization, extraction). The size of the reactor is usually not more than 1 m^3. Apart from the permanently installed secondary heating/cooling circulation, compact temperature-regulation equipment available as modules can also be used for heating and cooling. The mounted column, at only a few meters high, is not really suitable for more than the simplest thermal separations (e.g., the removal of solvents). The column usually consists of glass sections with random ceramic packing, but can be improved with the installation of structured packing consisting of high-quality corrosion-resistant specialized materials (e.g., Hastelloy).

A module is typically compactly built and comes in a transportable or, usually, mobile frame. Coupling systems are especially suitable for the attachment of the hosing connecting the different pieces of equipment, so that no tools are needed to connect and disconnect such links (see also Section 5.3). Alternatively, universal flanges can be used, also for connections to glass apparatus. Electrical connec-

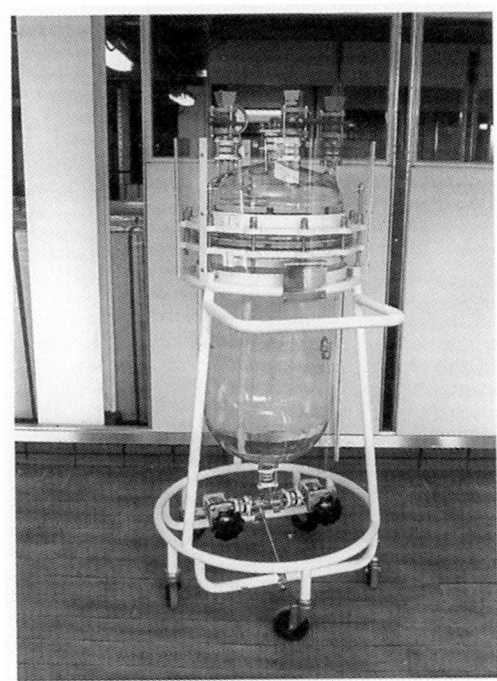

Fig. 3.12 A mobile glass vessel module

tions, for example, for the pump of a scrubbing tower module or for a vacuum module, are available as explosion-proof electrical plugs at the nearest plug socket (in explosive areas, where explosive vapor–air mixtures may be generated).

Typical modules are, for example, mobile glass vessels (Fig. 3.12), which may also be equipped with heating/cooling coils or stirrers. They may serve as receivers, loading vessels, or buffer containers for the handling of feedstocks, solvents, and products. For the evacuation of modular multiproduct plants, mobile vacuum aggregates, from simple liquid ring vacuum pumps to three-stage pump combinations (two steam-jet vacuum pumps and a liquid ring vacuum pump) in explosion-proof form, can be used. Scrubbing towers and jet scrubbers are available as modules for waste-gas scrubbing. Mobile filter modules, from a simple pressure nutsch (Fig. 3.13) to agitated-, swivelling-, and heatable filter driers (also known as process filters or agitated pressure nutsches) (Fig. 3.14) take care of solid/liquid separations. Of these, the filter drier is a special case in that it is a multipurpose module (see also Section 3.6); it can be used for several technical operations: filling and discharging under closed conditions, filtration, washing, and drying, and within limits: reactions, precipitations, and crystallizations. The product-handling abilities of the filter drier, operating in a closed system, have advantages over simple pressure nutsch filters that need to be discharged by hand while open: product transfer between the different process steps or even product campaigns and the resulting cross-contamination of products are avoided. This requirement is particularly important in, for example, the pharmaceutical, fine-chemicals, and

Fig. 3.13 Filter module: pressure nutsch. Reproduced with kind permission of KHS

Fig. 3.14 Filter module: filter drier. Reproduced with kind permission of KHS

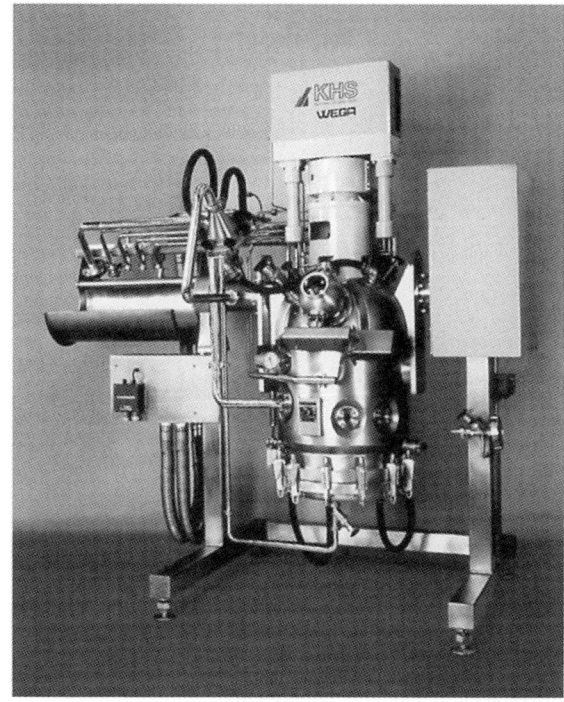

food sectors. With such filters, not only is possible product contamination avoided, but the danger posed by toxic substances or solvents to the operators is also avoided.

Other examples of modules are operational pump stations, used for bringing about forced circulation by a recirculation cycle or as dosing units. Modules can also be used for loading and dosing of solids; examples are mobile discharging stations for containers (Fig. 3.15) or for big bags, with an integrated weighing unit (weighing cells) and dosing equipment (e.g., screw-dosing apparatus).

In contrast to the situation in a conventional plant, where the corrosive properties of the substances that are used are known, the chemical resistance of the materials used in a multiproduct plant needs to be as extensive as possible (see also Chapter 6). For this reason, widespread use is made of glass-lined steel, steel/PTFE, glass, stainless steel, and specialized metals such as Hastelloy in modular multiproduct plants. The tubing and seals used in the areas where the products are handled are mainly made from materials based on PTFE and graphite.

The measuring and regulating instruments are standardized and generally correspond to the equipment of a conventional batch plant, but with the measuring and operating ranges as broad as possible (see also Chapter 7). The modular multiproduct plant is controlled in place and also over a distributed control system. The equipment for technical safety (see also Chapter 10) is also standardized and corresponds to that of conventional batch plants.

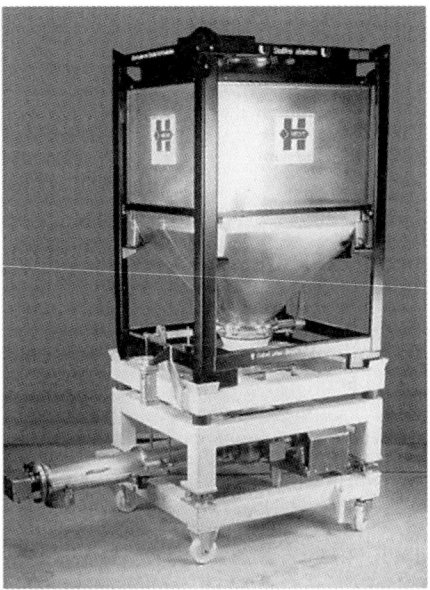

Fig. 3.15 Mobile station for discharging containers, with weighing unit and dosing apparatus. Reproduced with kind permission of Hecht

3.3.3
Application Areas and Limitations

Because the modules need to be mobile, the apparatus can normally not be larger than $1 \, m^3$. Since the agitated reactor vessel, its periphery, and the modules grouped around it only need approximately $20–30 \, m^2$ of space, construction in compartments with separate air supply and exhaust systems is possible; this makes not only the broad product spectrum, but also the safe handling of toxic substances possible. The operating temperatures are between $-15 \, °C$ and $+200 \, °C$. Because glass parts are used, the operating pressures vary from vacuum to low pressure (usually 1.5 bar). The many connection possibilities lend high structural flexibility (see Section 1.2) to this type of plant; its modular nature also makes it suitable for complex process-technological tasks. The capacity flexibility is low as a result of the limited equipment volume, and the product-assortment flexibility is middling.

A modular multiproduct plant has the following advantages over a conventional plant:

- Reduced costs (time and finances) for planning and constructing the necessary plant from developed, standardized modules
- High structural flexibility, therefore quickly adapted to changed or new requirements by the exchange of modules or use of additional modules
- Equipment (module) is used to full capacity: when the product campaign is over, the modules of the operating plant are available for the next product or process. Temporarily unused modules are kept ready in a pool.

The operator of such a plant is able to respond flexibly, rapidly, and economically to wishes of clients, also where complicated process engineering is involved. The clients are mainly from the research and development areas of the chemical industry, since modular multiproduct plants are located mainly in pilot plants, due to the limited size of the apparatus. There they serve as plants for pilot-scale process development and the production of sample- and market-introduction quantities of product (see also Section 2.2).

3.4
Multiproduct Plants with Pipeline Manifolds

3.4.1
Introduction

The concept of a multiproduct plant with pipeline manifolds is one that has been realized in many possible variations. With pipeline manifolds, the flexibility of various combinations of apparatus and machinery is enhanced to such an extent that the requirements of a multiproduct plant are easily met. This concept is ideal for developing multiproduct plants further, especially with regard to upgrading at a later stage and expansion of plants that were originally planned as monoplants. In this way, even established structures can afterwards be flexibly connected through the installation of pipeline manifolds.

A major advantage of this concept is that expensive special equipment is intensively used and need not be available as several pieces.

The costs associated with extra pipelines and manifolds are comparatively low. However, for this type of plant, the possibilities for cleaning the pipeline networks must be carefully thought out. For the upgrading of existing plants, the introduction of new pipelines, at least in part, needs to be considered.

3.4.2
Plant Structure

The basic construction of a multiproduct plant with pipeline manifolds is shown in Fig. 3.16.

For this type of plant, a variety of possible pipeline connections between the apparatus is already available. At junctions – the pipeline manifolds – the pipelines running between apparatus are joined. There the apparatus can quickly be connected through adapters to other facilities [3.19].

There are commercially available systems for the technical setup of this concept; these range from manually operated rapid-action couplings for pipes and pipeline adapters (see Fig. 3.17) to automatic manifolds. Fig. 3.18 shows such an automatic, piggable manifold with twelve inlets and four outlets. To circumvent the inherent danger of confusion between the different lines, interlocking systems

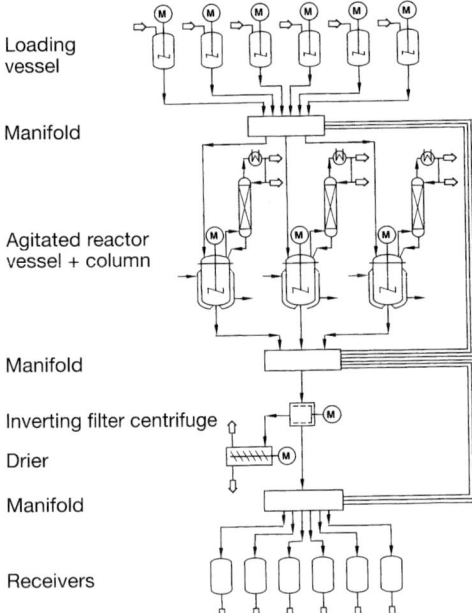

Loading vessel

Manifold

Agitated reactor vessel + column

Manifold

Inverting filter centrifuge

Drier

Manifold

Receivers

Fig. 3.16 Multiproduct plant with pipeline manifolds

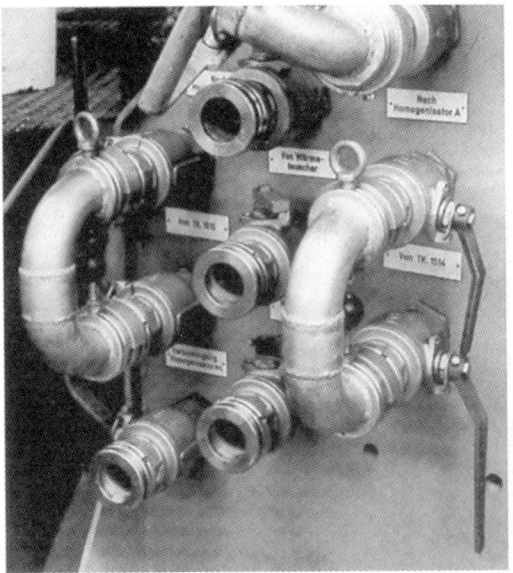

Fig. 3.17 Pipeline manifold with adapters. Reproduced with kind permission of I.S.T.

Fig. 3.18 Automatic, piggable pipeline manifold with twelve inlets and four outlets. Reproduced with kind permission of I.S.T.

Fig. 3.19 Indicator system for indicating the actual connections. Reproduced with kind permission of Balluf and Mollet

(see also Section 5.6) as well as indicator systems to indicate the actual connection are available [3.20] (see Fig. 3.19).

For biotechnology applications, purity and especially sterility is of prime importance. These requirements are easily met in multiproduct plants, for example by the use of sterilizable pipeline manifolds.

3.4.3
Application Areas

The operating conditions of pipeline manifolds are strongly determined by the couplings used (see Section 5.3).

Temperatures up to 150 °C and operating pressures up to 25 bar are possible. The operating pressures of the connected apparatus are, however, often limited to 3 bar; higher pressures are, nevertheless, possible in product-oriented (process-oriented) multiproduct plants.

It needs to be easy to pump the conveyed materials around, and there must be suitable solvents or other methods such as pigging systems (see also Section 5.5) available for purifying the pipelines.

Operating with flammable or toxic substances is only possible with self-locking and leak-proof couplings.

High structural and capacity flexibility is attainable with multiproduct plants with pipeline manifolds.

3.4.4
Variants of the Discontinuously Operated Standard Multiproduct Plant

A variation on the discontinuously operated standard multiproduct plant is a combination with pipeline manifolds; then, for example, a cascade of agitated reactor vessels could be used instead of the central agitated reactor vessel. The sizes of these vessels could be logically graduated, for example, 2 m³, 6 m³, and 12 m³. Appropriate relative positioning of the vessels would allow product flow by gravity. Not every vessel would be equipped with a distillation column or reflux condenser. The unit operations carried out in an agitated reactor vessel (Fig. 3.2) is then spread over different vessels.

Depending on the processes to be carried out, not all vessels would be in use. Appropriate pipeline circuitry lends the connections to the receivers the necessary flexibility.

A mixing vessel with mounted column, as described above, is not suitable for difficult distillations. The performance of a mixing vessel as evaporator is limited and not very product-friendly. The availability of a range of continuous and discontinuous distillation columns, which may be connected to the reaction vessels and their receivers via pipeline manifolds, is therefore a flexible solution for distillations in industrial complexes with standard multiproduct plants. Completely corrosion-resistant columns with random glass packing or structured ceramic packing are often available. For metal columns, stainless steel is employed more often than the more expensive materials based on nickel. Falling-film evaporators are used most often, but thin-layer evaporators (Sambays) are also widely applied.

For logistical reasons, multiproduct plants, especially when used for the production of a family of products, are connected to tank farms. Instead of receivers integrated into the plant, flexible connections to storage tanks or, for completed products, to filling stations are used, also for the loading of products. These are realized over pipeline manifolds, including piggable ones.

3.5
Pipeless Plants

3.5.1
Introduction

Over the last few years discussions in technical circles have focused more and more on a type of plant conceptually very different from the multiproduct plants discussed above, namely the pipeless plant. This type of plant has the distinctive feature that materials in it are not conveyed in pipes, but in mobile vessels, which move between the different stations where the different process steps are carried out.

The sequence of the required process steps thus determine the travel route of the mobile vessels. After a sequence has been completed, the mobile unit is cleaned in a specific station, while another vessel has already started going through the same or a different sequence.

This is practically the technical-scale version of the procedure followed by the preparative chemist in the laboratory, who, with glass flasks in hand, mixes the various chemicals together where they are found in different parts of the laboratory. The reaction of the starting materials then takes place in the fume hood, and the reaction flask is afterwards cleaned at the sink.

The technical realization of this concept is also not new: coatings factories have been operating as "pipeless plants" since the early 1980s. The concept has, however, been further developed and widely automated by Japanese engineering firms [3.21].

In pipeless plants the CIM (computer integrated manufacturing) concept is almost completely realized, so that a closer look at this type of operation is justified, as this type of plant has obviously become established in the Japanese coatings industry; by now there are supposed to be more than 20 "pipeless plants" in operation.

3.5.2
Plant Structure

In contrast to a conventional batch plant, where the reactor is installed in a fixed position, the principal equipment in a pipeless plant is mobile (Fig. 3.20). The mobility of the main equipment is associated with several advantages:

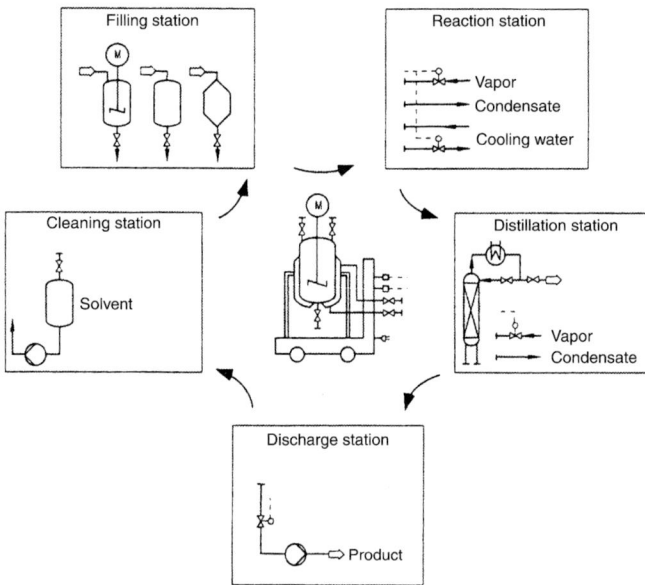

Fig. 3.20 Pipeless plant [3.22]

- The order in which the process operations are carried out can easily be modified by changing the route between the different stations
- Several different units of the primary equipment, also of different sizes, can be operated at the same time
- Substances are not transported through pipelines from one station of the technical process to another; this reduces the problem of cross-contamination, as well as contamination by previously utilized substances; the problem of cleaning between different products is also reduced

Since manual work, such as cleaning of pipelines and equipment, is considerably reduced with this type of plant, a high degree of automation is possible.

For the practical application of this concept, technical solutions need to be found for operations that are still borrowed from processing technology, such as positioning systems for the exact docking of mobile vessels at the stations, or the tools for production planning (see also Chapter 9).

The fundamental component of a pipeless plant is the transport system responsible for conveying the product flow between the stations. There are several competing systems, as the following examples show:

- Two-dimensional systems (Fig. 3.21):

Rail-transported systems, available as horizontal (AIBOS 5000, ASAHI Engineering Co.; MILOX, TOYO Engineering Corp.) and suspended (spacetram, Kuraray Engineering Co.) varieties.

Fig. 3.21 AIBOS 5000, ASAHI Engineering Co.

Self-transporting systems, on wheels (AIBOS 8000, ASAHI Engineering Co.; MI-FLEX, Mitsui Engineering and Shipbuilding Co.) and air-cushioned, have also been developed.

- Three-dimensional systems:

These have been developed out of high-shelf stores (FMS, Mitsubishi Kasei Engineering Co.).

- Rotating systems (Fig. 3.22):

The central equipment (reactor) is mounted on a vertical axis and, by rotating, can direct the operations, such as filling and emptying, of various pieces of equipment (MULTIMIX, Mitsubishi Heavy Industries Ltd.).

All the systems mentioned above are fully automated. Transport of substances as well as docking in and out at the fixed stations take place without manual intervention.

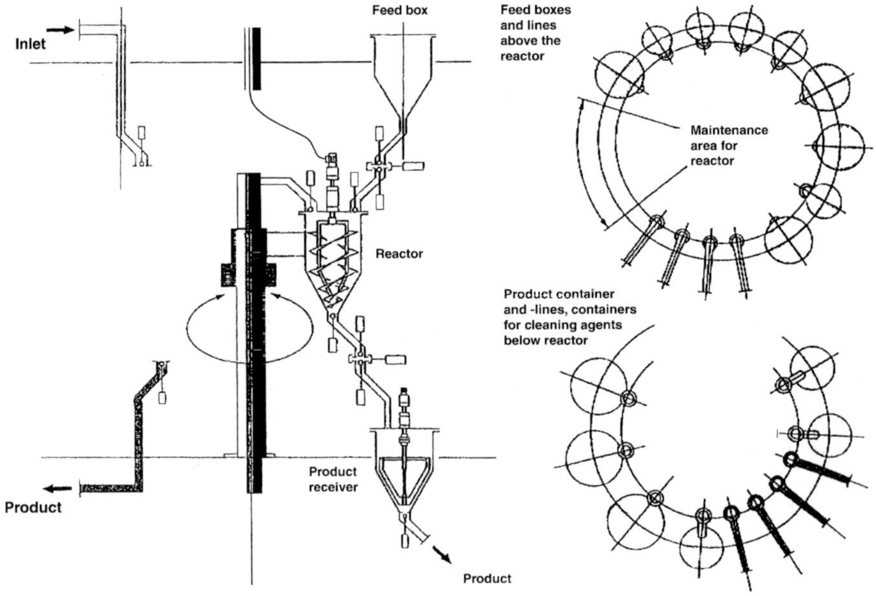

Fig. 3.22 MULTIMIX, Mitsubishi Heavy Industries Ltd.

A requirement for automated running is a coupling system that enables fault-less connections between the mobile central unit and the various fixed stations. Connections between pipelines and to the power supply as well as the use of sig-nal transfer units are required for this (Fig. 3.23). It is important that the mobile units of the pipeless plant are precisely positioned; that vessels of different sizes should be connectable adds to the difficulties. The requirement that connections should be leak-proof is just as valid here as in fixed pipeline connections.

Connections with inflatable seals are used for connecting pipelines that convey solids. Flammable, toxic liquids can also be handled, as long as the couplings are pocket-free, provided with rinsing systems, and self-closing. Technical solutions for these problems are already available, but still need to be adapted to pipeless-plant technology.

The high degree of automation achievable in pipeless plants makes the use of production planning and process control systems especially useful. For this, exten-sive sensory equipment is essential, to report on the process engineering condi-tion of each piece of mobile equipment and of the fixed stations, as well as the location of each mobile unit with regard to space and time. If optimization pro-grams are combined with this, production can be controlled so flexibly that the plant can run to full capacity, and storage quantities can be minimized.

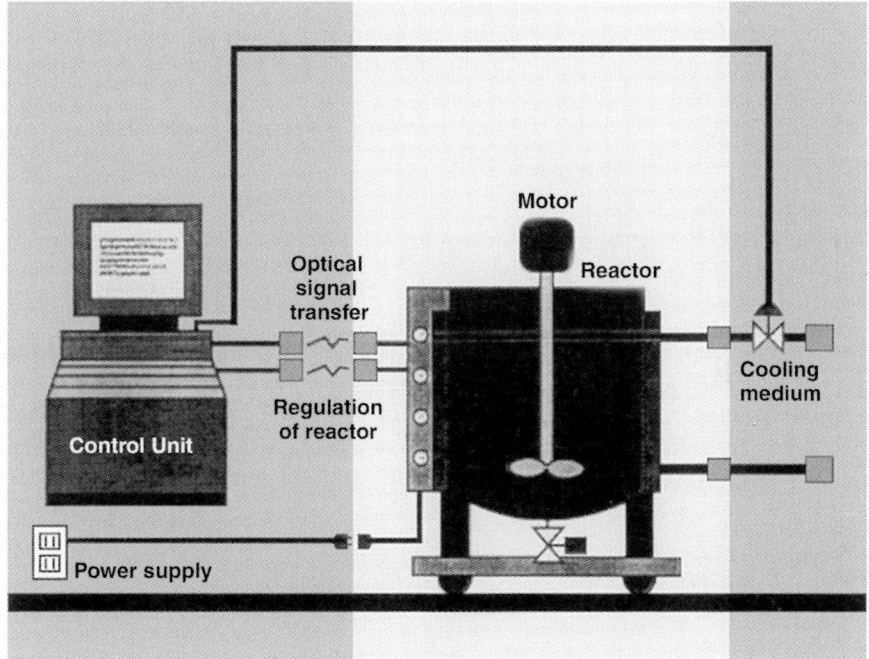

Fig. 3.23 Mobile central unit, ASAHI Engineering Co.

3.5.3
Application Areas

When very many feedstocks and formulations are used or when the transport of substances through pipelines is very difficult, that is, when logistical problems are dominant, pipeless plants provide a very good technical solution [3.23].

Pipeless plants are used in the Japanese coatings industry, and it is only a matter of time until this technology is also utilized in the Western industrial nations to a broader extent. There have also been reports on the use of this type of plant in the production of lubrication oil and adhesives. Up to now, there have been no examples of its commercial use in the chemical industry. Why is it so?

So far, the process engineering complexity of the processes carried out in pipeless plants has been low. The process steps of these units are often limited to loading, dosing, mixing, stirring, discharging, and cleaning.

There are several reasons why this process is limited to these steps:

The conceptual advantages of pipeless plants are best realized when all process steps can be carried out in one vessel. Even when stations for distillation, for example, can be envisaged, the advantage of easy cleaning is lost, since not only the mobile vessel, but also the distillation station needs to be cleaned at product changeover.

For reactions to be carried out, it is essential that transport takes place in closed containers and that it is possible to operate equipment under pressure. Such technical requirements for the safe running of reactors are still beyond the concepts available at the moment for pipeless plants.

Reaction stations, with the facilities required for carrying out reactions, are being developed at the moment. Examples are:

- Agitation units
- Heating/cooling circulation units
- Safety units for release of pressure
- Dosing systems for reactants

Developing technical concepts and detailed solutions to extend the use of pipeless plants to the chemical industry is an interesting new area of endeavor.

In this context it should also be clarified to what extent the available technical regulations are suitable for the authorization of a plant of this type. For this, interpretation work needs to be done in conjunction with the authorities.

3.6
Plants with Multipurpose Apparatus

Plants with multipurpose apparatus characteristically operate discontinuously; a distinguishing feature of these types of plants or apparatus is that several unit operations follow each other in a single piece of universal apparatus. This is in contrast to discontinuously operated standard multiproduct plants, where the number and sequence of the unit operations are generally fixed beforehand as a result of the available apparatus and the interconnections.

The universal apparatus is typically only a further developed form of commonly used and well-established apparatus applied in solid processes. According to the definitions given in Section 1.2, these are product-oriented multiproduct plants for the production of a palette of small quantities of products. A limited process- or product-assortment flexibility is possible if the various available unit operations are made use of in different ways. There is generally also a technologically oriented component involved in the choice of a multipurpose apparatus concept, to ensure a broad application area, expressed in large investments, and to ensure support for the planned products with similar or not yet closely defined properties.

In what follows, the term "universal apparatus" will be consistently used for this type of apparatus, although the universal applicability of these apparatus may be doubtful, both with regard to its implementation as well as concrete individual examples.

In a sequence of basic operations, mainly associated with liquid processes (although, for example, limited formation and handling of suspensions is also possible), the proven and most widely used type of universal apparatus is the agitated vessel, often combined with receivers, dosing stations, mounted columns, condensers, phase separators, external heating/cooling circulations, fraction collectors, and so forth. The agitated vessel is, however, seldom found in the same category as the multipurpose apparatus dealt with in this section, since the apparatus described below mainly originated in solids processing.

In practice, reaction and phase separation are often combined in one agitated vessel, or the same vessel is used afterwards as receiver or liquid-phase evaporator for a mounted column. More common, however, is the use of agitated-vessel methods in a combination of synthesis and purification steps, such as solvent removal by distillation and/or workup of the product, and then these steps are carried out in several different apparatus. The agitated vessel as universal apparatus will not be dealt with further in this section, as this subject has already been covered in detail in Section 3.1, on discontinuously operated standard multiproduct plants.

Other universal apparatus from the area of fluidized processes, such as jet reactors, bubble-column reactors, and so forth, are less important here.

A clearly larger number of variations of universally applicable discontinuously operated multipurpose apparatus have been developed in the area of solid-process technology. The development of many variations may at first appear contradictory, as one may suspect that these multipurpose apparatus are not universally applicable. The basic reason for this is, however, that the requirements are much more diverse and complex than in the corresponding fluid technologies. This also means that there are only limited possibilities for covering broader application areas in solid-process technology and that this is also associated with more effort and correspondingly higher costs.

In almost all known application areas, the final product is a solid. These universally applicable apparatus are mostly further developed forms of known and proven basic types normally used in the basic operations, and originate exclusively from solid-process technology (e.g., reference [3.24]).

Two lines of development are recognizable:
- Apparatus only capable of handling *preformulated solids* (e.g., fluidizable product for fluidized-bed reactor)
- Apparatus handling *solids* in virtually *any* form

The need for a preformulated solid understandably hinders wider application most, as product-specific auxiliary units are usually required for preformulation. The number of applications has therefore remained limited, and this type of apparatus will therefore not be discussed further.

A broad variety of apparatus has, on the other hand, been developed from the second line of solid-process universal apparatus. Further development of well-established basic forms such as shovel driers, kneaders, conic mixers, or agitated pressure nutsch filters are recognizable in universal or multipurpose apparatus that have, in the most various ways, become established in practice.

3.6.1
Multipurpose Apparatus with Design Based On Shovel Driers

- Shovel driers with modified drive to handle more viscous phases (from standard providers of shovel driers). Special developments such as:

- Tiltable solid-process reactor, produced by Bertrams
- Inox universal drier, produced by InoxGlatt

The background to this development is that, due to their design, many of the old generation of shovel driers in the typical operating sizes have drives that are too weak (power $\leq 3\,kW\,m^{-3}$), operating at too many revolutions (5–8 min^{-1}). This leads, already with relatively harmless products, to damage to the shafts and gears. A well-proven remedy in the case of products with viscous phases was to use shovel driers with beater bars. But the significant step was the changeover to hydraulic-driven (or similarly driven) driers, capable of achieving higher torque (power typically 5–10 kW m^{-3}) at a lower number of revolutions (2–5 min^{-1}) and equipped with inherent protection against system overload.

Another criticism of conventional shovel driers was the relatively long times required for discharge. A first step towards reducing this was to change the rotational direction of the rotary blades. This led to shovel elements that, depending on the rotational direction, could bring about different transport effects, leading to discharge taking place centrally or at the front. But even this solution can lead, with frequent product changeovers, to product loss (with all the resultant problems such as cleaning with salt, flaked ice, etc., waste-product disposal or repeated workup), since the area between the shovel element and the wall is essentially dead space where the product is not forced to move about. This led to a special line of tiltable apparatus, for easy and fast discharging, becoming available (e.g., tiltable solid-process reactor of Bertrams, Fig. 3.24).

A further special development was the integration of a chopping shaft into the actual stirrer of the shovel reactor (Inox universal drier of InoxGlatt, Fig. 3.25), as a way of handling viscous or lumpy product phases.

All the above examples of multipurpose apparatus show how minor modifications could significantly extend the application areas, so that a broader range of product consistencies can be handled.

- Fast-running shovel drier with built-in mills, choppers, etc., as required (Druvatherm of Lödige; Turbulent reactor of Drais)

This type of shovel drier represents a combination of fast-running mixer and conventional shovel drier. The main aim is economic handling of the product during the different basic operations. The centrifugal forces generated by the shovel elements result in a considerably larger heating surface coming into contact with the product, making more heat available for the reaction or for drying, as the product is constantly thrown out of the bulk and against the walls, in contrast to conventional shovel driers, where the product mainly stays at rest at the bottom of the drier. The more intensive mixing and the thorough contact between product and walls result in considerably faster heating, cooling, or drying. As a result, much shorter time periods are required compared to conventional shovel driers. This type of apparatus is also characterized by low-resistance shovel elements (plowshare or related shapes), helping to reduce the required operating power. The typical operating power is between 10 to 15 kW m^{-3}, the characteristic number of re-

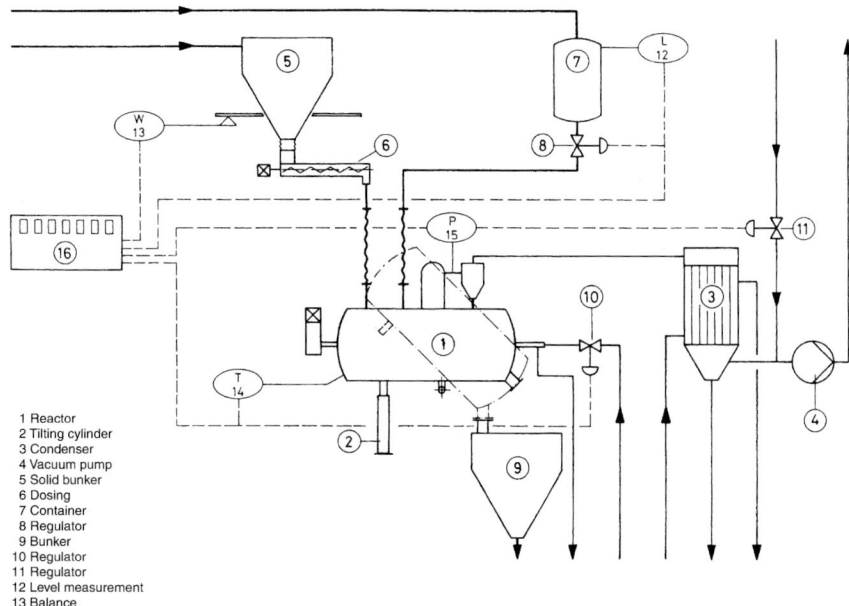

1 Reactor
2 Tilting cylinder
3 Condenser
4 Vacuum pump
5 Solid bunker
6 Dosing
7 Container
8 Regulator
9 Bunker
10 Regulator
11 Regulator
12 Level measurement
13 Balance
14 Temperature measurement
15 Pressure measurement
16 Control unit

Fig. 3.24   Tiltable solid-process reactor. Reproduced with kind permission of Bertrams

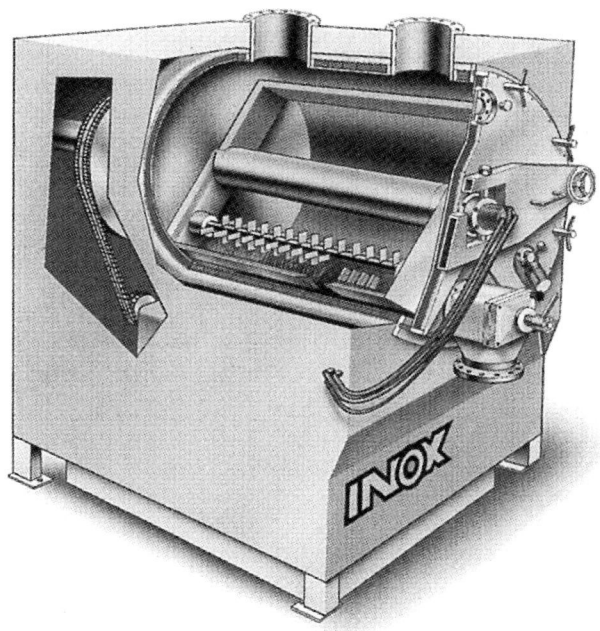

Fig. 3.25   Inox universal drier. Reproduced with kind permission of InoxGlatt

Fig. 3.26 Druvatherm. Reproduced with kind permission of Lödige

volutions is 20–40 min^{-1}. These data show that the available torque may be lower than in the above shovel-drier version; this means that the handling of a wide variety of product consistencies is only possible at low revolutions and correspondingly high torque. The torque-dependent regulation of the number of revolutions makes adapting to the product consistency at hand possible.

Another step in the same direction is the use of one or more fast-running mills or choppers (Fig. 3.26: Druvatherm of Lödige), incorporated into the apparatus and operating in the space between the shovel elements. To ensure that the amount of wall surface that thereby no longer comes into contact with the shovel elements remains as small as possible, only the mill shaft is usually arranged between the shovel elements – in spatial terms, the actual chopping elements of the mill operate only in the areas beyond the shovel elements.

The importance of easy cleaning of the equipment was emphasized from the beginning of the development; this led to the first apparatus with mixers that are easily removed (Fig. 3.27: "Turbulent" reactor of Drais). Peripheral parts are usually similar to those of the slow-running shovel-drier types.

3.6.2
Multipurpose Apparatus with Design Based On Equipment for Highly Viscous Products

(Discotherm B batch, AP Reactor batch of List; Reactotherm RT batch of Krauss-Maffei)

The development of this type of apparatus was the result of the following operating problems which couldn't be resolved satisfactorily by the two previous shovel-drier types:

Fig. 3.27 "Turbulent" reactor. Reproduced with kind permission of Drais

- For extremely viscous products, both of the previous shovel-drier types were, despite all adaptations, only of limited use: their drives do not have enough power
- Highly viscous products tend to be adhesive; in the previously discussed types this leads to parts of the product sticking to the shafts, thereby not participating in the actual process
- With both of the two previously mentioned types of apparatus, the heat-transfer surfaces are generally limited to the casing of the apparatus. Heating of the blades is more a protective measure than an intentional part of the process. A significant increase in the heat-transfer surface per unit volume was no longer possible in the two concepts discussed above.

For these reasons, some firms initiated the development of what can be regarded as self-cleaning kneaders. They can be differentiated into single-shaft apparatus, with permanent kneading/cleaning elements (Fig. 3.28, Discotherm B batch of List, another similar model is Reactotherm RT of KraussMaffei) and double-shaft apparatus (all-phase reactor, or AP reactor batch of List), in which the meshing action of the shafts results in mutual cleaning. These designs belong to the most expensive chemical and engineering machinery and apparatus.

The first line of apparatus combines a high degree of self-cleaning with large heat-exchange surface area per unit volume (Discotherm version). This is achieved with propeller-like disk-shaped segments situated equidistantly along the shaft.

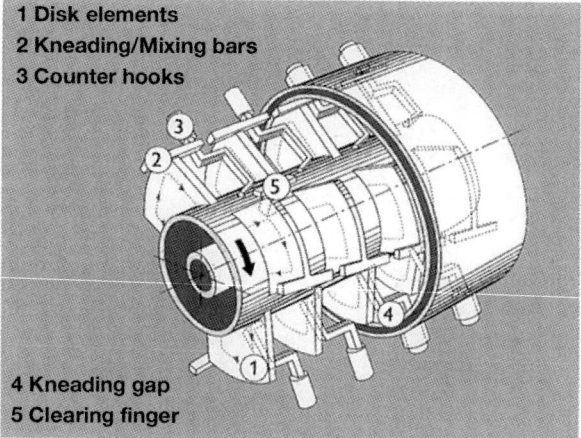

1 Disk elements
2 Kneading/Mixing bars
3 Counter hooks
4 Kneading gap
5 Clearing finger

Fig. 3.28 Discotherm B Batch. Reproduced with kind permission of List

The kneading mechanism is brought about by the kneading bars at the end of each disk segment, while the stationary hook elements that protrude into the spaces between are responsible for the cleaning. Cleaning of such an apparatus is difficult, because of the many hooks and kneading bars. For the same reasons, the shaft can only be moved once it is in a specific position. The normal operating power is 20–40 kW m^{-3}, or more as the apparatus goes more in the direction of being a kneader.

This type of highly specialized equipment easily costs twice or three times as much as a normal shovel drier. The manufacturing costs make up a significant part of the expense: the significantly reduced distance to the walls of this apparatus compared to conventional shovel driers, with the associated enhanced heat exchange, is the result of its solid build. Because it is highly specialized, this type of apparatus is seldomly used as multipurpose equipment, although its design and facilities would make it well suited for this purpose; it is mainly used in single-product plants.

For the sake of completeness, the double-shaft apparatus (e.g., all-phase reactor, or AP reactor of List) is also mentioned here, although its importance as universally applicable equipment is only marginal. This is not only the result of its high cost, it is also because of the reduction in the available volume with the development of this apparatus in the kneader direction. This means that it cannot be used in basic operations with large volume variations.

3.6.3
Multipurpose Apparatus with Design Based On Conic Mixers

(Mixer-drier MT of KraussMaffei, Vrieco-Nauta mixer of Hosokawa Micron, conic mixer integrated into the plant unit: the Titus system of KraussMaffei)

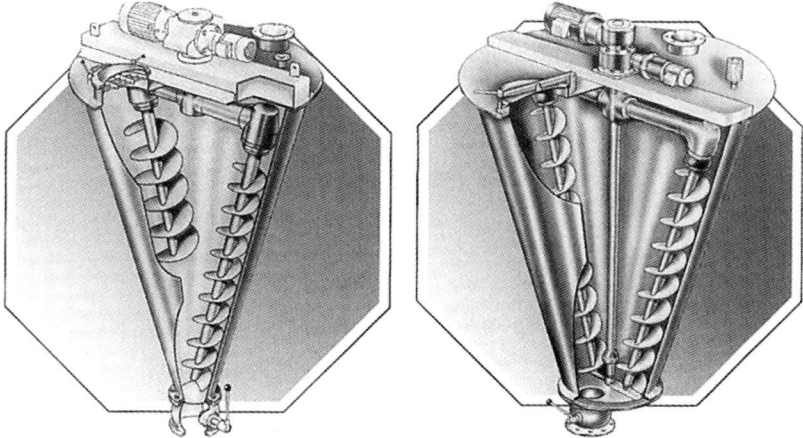

Fig. 3.29 Conic mixer. Reproduced with kind permission of Hosokawa Micron

A further direction in the development of multipurpose apparatus concerned the design of conic mixers, to make them more widely applicable, by the addition of features such as heatable and coolable walls or additional filter or chopping elements. There is, however, little known about the application of this system as a universal chemical reactor, probably because it is not really designed for viscous phases.

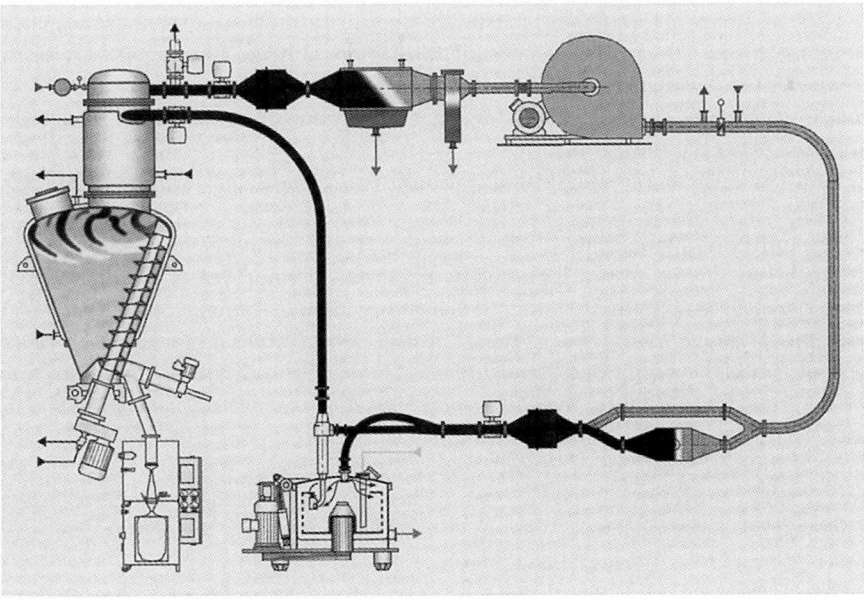

Fig. 3.30 Titus system. Reproduced with kind permission of KraussMaffei

Hosokawa Micron provides the greatest variety of conic mixers (e.g., Fig. 3.29), apart from also supplying standard lines with Vrieco-Nauta ribbon mixers, satellite screw mixers, double-screw mixers, combi-mixers, and so forth.

The Titus system of KraussMaffei (Fig. 3.30) is universally applicable, and also has a conic mixer–drier as central component. This system is preferred in the areas of fine chemicals and pharmaceuticals for the isolation of solids by its combination of filtration and drying processes. These do, however, take place in two separate apparatus: the mentioned conic mixer–drier and a vertical centrifuge.

3.6.4
Multipurpose Apparatus with Design Based On Agitated Pressure Nutsch Filters or Nutsch-Filter Driers

(Nutrex of Rosenmund, Innopro Wega filter drier of KHS Klöckner-Holstein-Seitz)

The last type of multipurpose apparatus system discussed in this context is that based on a further development of agitated or mechanically dischargeable pressure nutsch filters. The best-known system is, without doubt, Nutrex of the company Rosenmund (Fig. 3.31); it allows the combination of the following steps: reaction, crystallization, extraction, filtration, washing, and drying [3.25, 3.26]. The Nutrex basically consists of a swivel-mounted closed cylindrical container with, on the one side, the reaction and drying part (in principle an agitated vessel), and, on the other side, the filtration part (based on an agitated pressure nutsch filter).

The Innopro Wega apparatus of KHS Klöckner-Holstein-Seitz (Fig. 3.32) is a similar system, in which any required combination of the above-mentioned steps is also possible.

These systems are available as standard chemistry models or as pharmaceutical versions; in the latter case, the design is adapted to make the preparation of extremely pure substances possible, which means that the apparatus should be highly cleanable, meeting the GMP (good manufacturing practice) guidelines, amongst other requirements.

Both systems can be equipped with a number of additional features, making it possible for a complete unit to come from a single supplier.

The universal applicability of all the multipurpose apparatus discussed above is also achieved by the use of high-quality materials or improved or additional facilities such as built-in mills, additional heating/cooling elements at the shafts of the apparatus, the ability to operate at unusual temperature ranges, the incorporation of facilities making operation under vacuum and under pressure possible, as well as good cleanability, facilitating easy product changeover, for example, in models with removable mixing elements, or the incorporation of comprehensive sampling units. Such multipurpose units may also be improved by diverse peripheral facilities, for example, dosing stations for liquids and solids, several receivers, mounted columns, cyclone separators or dust filters, vacuum aggregates or several collecting tanks for fractional distillation.

The application areas for these universal apparatus are, at first glance, huge. Two basic points must, however, be kept in mind:

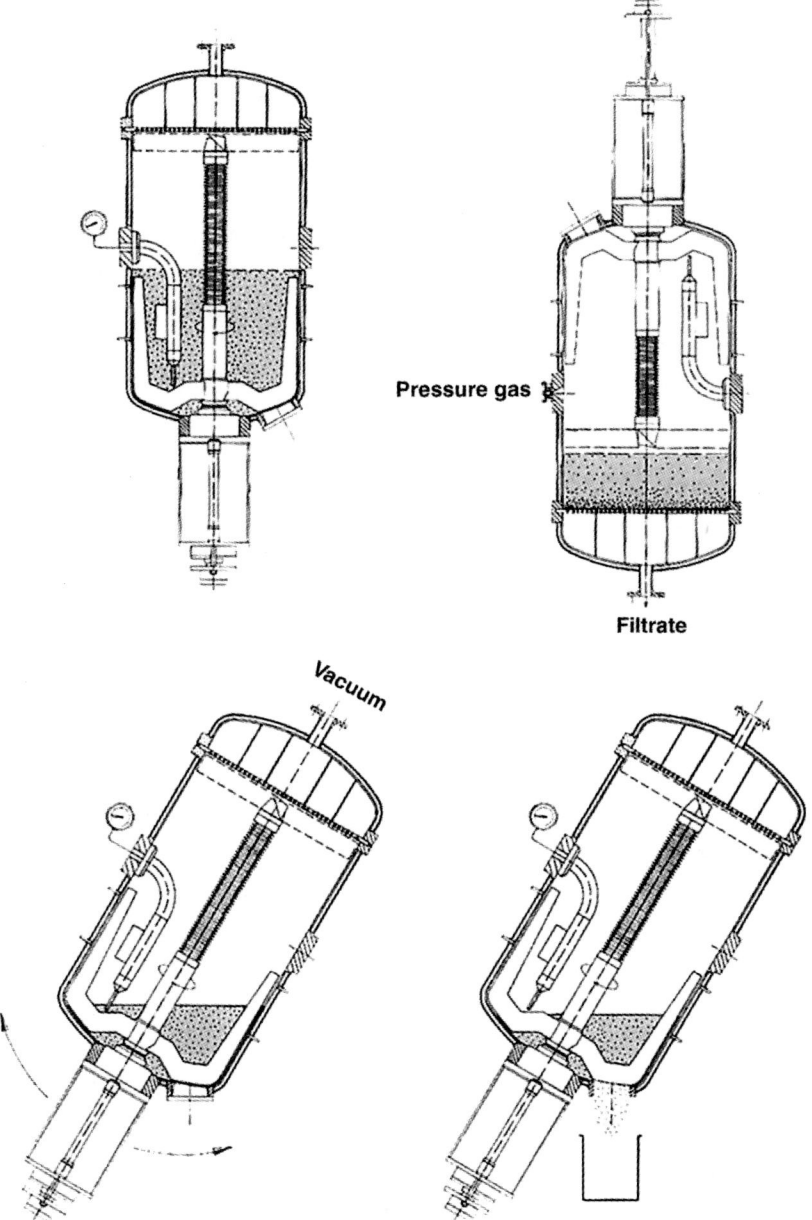

Fig. 3.31 Nutrex. Reproduced with kind permission of Rosenmund

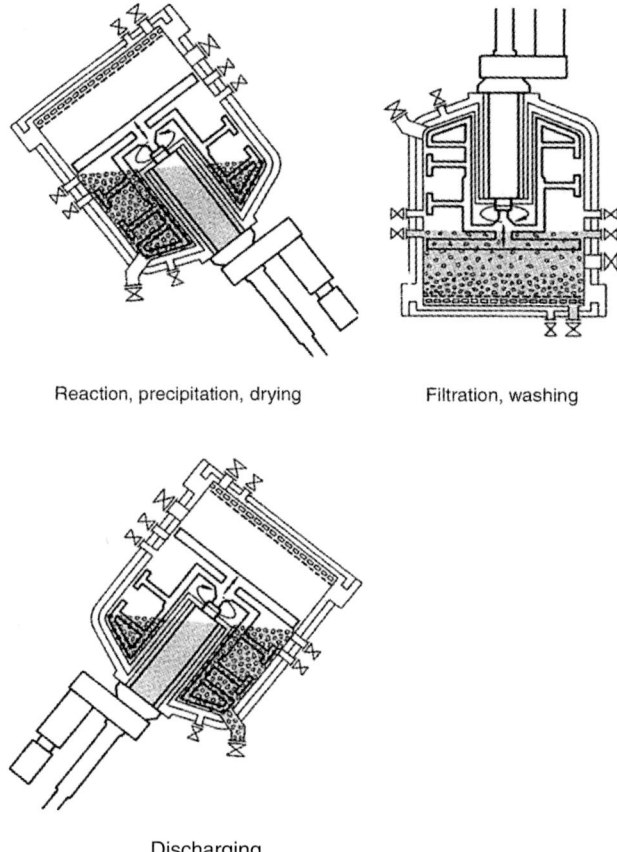

Reaction, precipitation, drying   Filtration, washing

Discharging

Fig. 3.32 The Innopro Wega apparatus. Reproduced with kind permission of KHS Klöckner-Holstein-Seitz

- In a given sequence of unit operations, there may be large variations in volume. The individual steps need to be checked to ensure that the universal apparatus fulfills its function with both the smallest and the largest volumes. If not, provisional collectors need to be available at suitable points, some steps or sequences of steps need to be repeated a number of times, or the process needs to be changed.
- These universal equipment are often developed from standard apparatus, but with increased specialization for a specific function, for example, the handling of highly viscous phases. However, the increasing specialization of equipment results in a decrease in the number of products requiring this specialization; this makes economically feasible operation or operating at full capacity dubious. In such a situation, if only one product falls away, the implications to the costs can be severe. The search for an economically viable replacement product is often very time-consuming, because the production of a new product needs to be

adapted to the available system, and only minor modifications to apparatus are usually possible.

The main areas in which these equipment are used are in solid/solid, liquid/solid, and gas/solid reactions, as in the synthesis of dyes and pigments. Other applications are in the pharmaceutical, foodstuffs, and pesticides sectors, where combinations of basic operations excluding synthesis are often used.

The manufacturer of such multipurpose apparatus is often faced with a difficult situation:

- On the one hand, the design of the apparatus has to be more complex than that of the conventional basic version, for dealing with the more critical material properties of the products
- On the other hand, the apparatus should not impede smooth functioning and should not be associated with large expenses, because in that case, products suitable for the process will not be available, and the few that can be produced may be associated with very high running costs. For many products, alternative, simpler, and therefore less expensive systems are then available.

The process engineering design of such universally applicable solid-process apparatus is usually based on experiments that have been scaled up to plant scale with suitable apparatus-dimensioning programs. Individual steps such as discontinuous distillation can be modeled in known ways. The planning of the various product syntheses encompasses not only considering cycles and logistical questions, but also implementation in a discontinuously operated standard multiproduct plant.

3.7
Peripheral Equipment of Multiproduct Plants

3.7.1
General

In this section, the apparatus and machinery generally found in multiproduct plants, but not dealt with under the concept of multiproduct plants (Sections 3.1 to 3.6) will all be described together under the designation of peripheral units. The peripheral units are only seldomly used for every process in a multiproduct plant. It may, however, be assumed that every product produced in the plant will need some part of these facilities.

In the chemical and related industries, the concept of the plant is used to describe the entirety of facilities, that is, apparatus and machinery, that make up the industrial complex. Usually just one large, continuously operated single-line plant makes up such an organizational unit, whereas a multiproduct industrial complex is usually made up of several multiproduct plants, in our sense of this term (see Chapter 1). Multiproduct plants are sometimes combined with monoplants into a commonly run industrial complex. Such plants are, apart from this, often combined into an integrated chemical site, where a whole range of industrial complexes cooperate closely. Especially multiproduct plants benefit from being placed at integrated chemical sites. Through the use of other resources, available on location, such as feedstocks and apparatus in other industrial complexes, as well as joint disposal facilities, the flexibility of the plant is enhanced.

It follows from this that the necessary peripheral facilities of multiproduct plants are closely joined to the rest of the works (site) in question. In this section, we will deal only with the peripheral equipment belonging to the individual industrial complex, not those belonging to the whole works (site), such as the cleaning plant, sewage works, and incineration plants.

If several multiproduct plants are to be combined into one industrial complex, the planning of the industrial complex should incorporate the required flexibility. Parts of the unit that belong together with regard to purpose or function should be placed together. Between the various functional parts of the industrial complex, room for transport and also extensions should be planned in. The individual lanes with reaction apparatus (each a multiproduct plant) should be placed together in one part of the building. In another part of the building (section), peripheral equipment, for example, for thermal separations, such as distillation and absorption columns [3.1] should be found. It is also advantageous to group all facilities that operate under GMP conditions into a separate part of the building. Other segments can be made up of exhaust-gas treatment facilities, formulation units, and stores. There should be generous space available between the different segments of the building for transportation and pipelines.

Fig. 3.33 shows the basic construction of a multiproduct industrial complex, according to reference [3.1]. The illustrated industrial complex has four discontinuously operated standard multiproduct plants available, a workup and loading lane for solid substances, as well as distillation and scrubbing units. The plants are connected to the associated tank farm and to each other through pipeline manifolds. Stores as well as water purification units and waste-burning furnaces are available for common use by all the industrial complexes on site (or within the works).

3.7.2
Peripheral Facilities for Individual Plants

Facilities for heating and cooling belong directly to a multiproduct plant. The heating–cooling circulation associated with an agitated reactor vessel as central apparatus iss described in Section 4.4.2. The same is valid for vacuum units, which often belong directly to a specific plant. In both cases, the use by more than one multiproduct plant or lane operating in parallel is conceivable. The simplest case is when the different plants never operate at the same time. Otherwise an economic optimum needs to be found. On the one hand, it is difficult to use peripheral facilities such measuring and regulation instruments or safety equipment for more than one plant. But on the other hand, multiple installations of vacuum pumps, for example, can be very expensive.

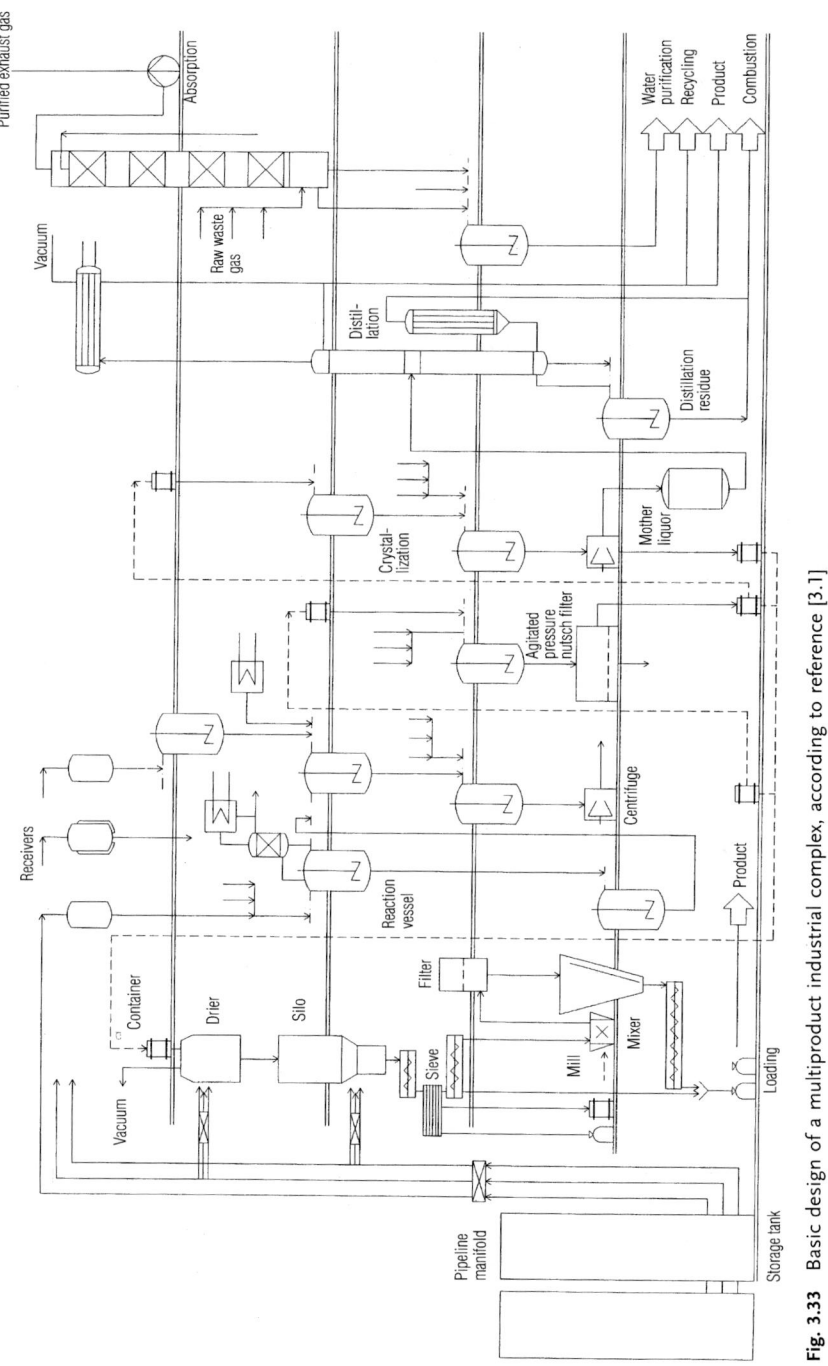

Fig. 3.33 Basic design of a multiproduct industrial complex, according to reference [3.1]

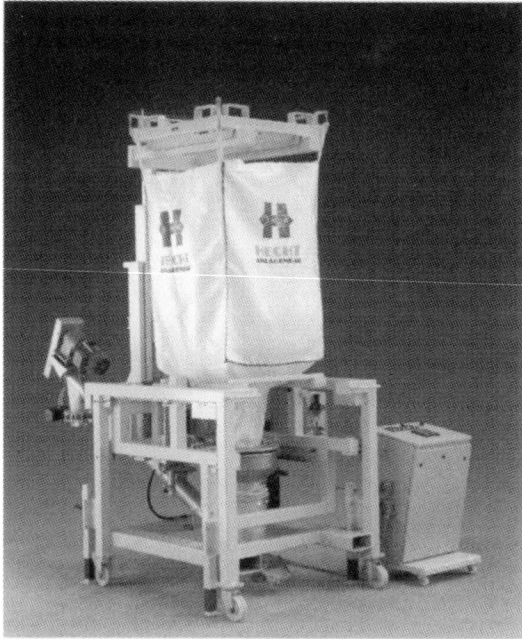

Fig. 3.34 Mobile emptying station with integrated balance, dosing screw, and control unit. Reproduced with kind permission of Hecht

Treatment of exhaust gas generated during the reaction stage is usually directly associated with the central process. The plant designer also needs to decide at this point whether the multiproduct plant should have its own exhaust gas treatment facilities. One may, for example, erect scrubbing columns, which can be operated in either acidic or basic form, in the near vicinity of the multiproduct plant. A complete exhaust gas module, commercially available (see Sections 4.4.2, Fig. 4.20), could, alternatively, be used. Depending on the requirements, such mobile modules can be operated wherever the corresponding exhaust gases are generated in the plant. Another example is a mobile station for the emptying of big bags or containers of solids. Modules completely equipped for dosaged filling in plants are also available on the market. One of these is a module with integrated balance cell, a solid-dosing screw, and corresponding controls with control unit. Fig. 3.34 shows a mobile big-bag emptying station with integrated balance. Such a module is used as a peripheral facility for only one plant at one time, but could in principle be available to the whole operation.

3.7.3
Peripheral Facilities for the Entire Operation

As described above, some multiproduct plants are found in association with other multiproduct plants, others are found together with monoplants within larger industrial complexes. It therefore makes sense to plan and operate peripheral facilities so that they can be used for the whole industrial complex. It is, for example, possible to use secondary heating and cooling circulation systems for an entire industrial complex, as is done for exhaust gas collectors and scrubbing units. Planning of multiproduct industrial complexes should, however, take certain incidental conditions into account. In contrast to monoplants, the peripheral facilities should also be very flexible. Alterations to these are also normal. During planning of the building, special care should be taken, where possible, to dedicate a separate part of the building to the peripheral facilities. This also requires the construction of pipeline routes that connect the multiproduct plants or the individual lanes within a multiproduct industrial complex to the peripheral units.

The waste-gas treatment units are, for example, well suited to being concentrated in one part of a building. The waste gas is easily collected through waste-gas collecting systems from the different parts of the multiproduct industrial complex, from where they can be directed to the central unit. Waste-gas flues in appropriate sizes can already be planned during design of the building, to connect all parts of the building [3.28]. Central are the scrubber towers or waste-gas incinerators for all the waste gas generated. Ventilators are installed on the clean-gas side (on the dust-laden gas side in incinerators) to maintain the vacuum in the waste-gas collectors. The vacuum in the waste-gas system is around 10–100 mbar.

A separate waste-gas collection system is also possible for connecting the safety installations against too high pressures; at the end of these, corresponding "catch tanks" are installed for collecting the contents blown out by the pressure tanks. Central waste-gas treatment systems are especially suitable for plants in which product families are produced, that is, where the processes are very similar, for example, phosgenation plants.

Similarly, facilities for the decentralized treatment of waste water in multiproduct production may also make sense. It is exactly because of the frequent product changeovers that the waste water should be dealt with very carefully. A buffer to catch waste water before its release into the sewage system can be used. In such a system, the waste water of the entire plant or a part of it can be saved in one of two buffers. If the first container is full, the second one is switched to. The first container is then emptied into the sewage system only after being analyzed and found acceptable. The waste-water route may also include appropriate solid/liquid separation equipment, such as filter presses, to separate insoluble components in the waste water.

For the energy required for heating and cooling of the various processes, the multiproduct industrial complex will generally be dependent on the network provided within the works. The supply of cooling brine is usually an exception to this. For this, an own cooling system with sufficient storage space for the coolant should be on hand. Glycol/water mixtures are well suited for cooling; this system easily facilitates the running of processes at temperatures as low as −15 °C. Fig. 3.35 shows the basic construction of a refrigerant system.

To avoid contamination of the cooling water, the construction of a secondary cooling cycle is recommended for the multiproduct industrial complex as a whole. In contrast to monoplants, where secondary cooling cycles are directly aimed at possible danger spots, flexibility in a multiproduct plant is only ensured if the

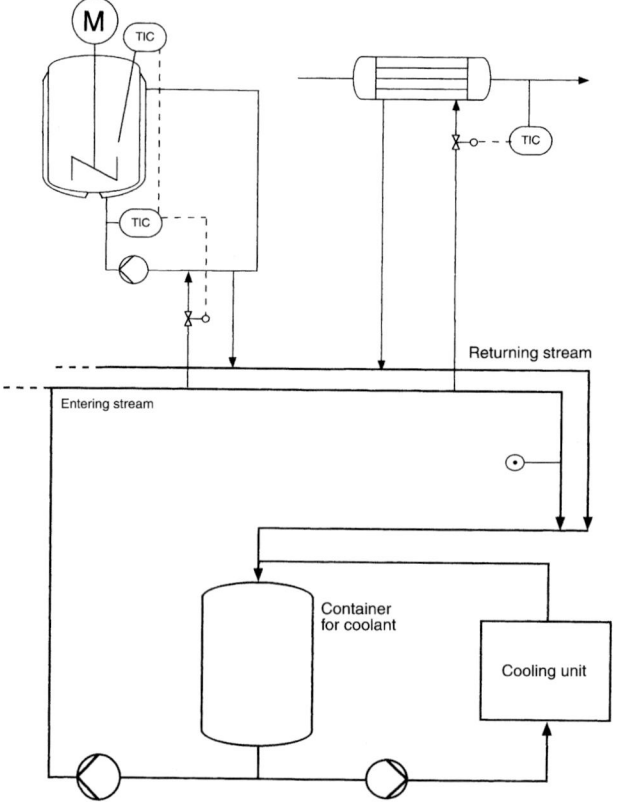

Fig. 3.35 Basic construction of a cooling-brine supply system

plant as a whole is secured. Such a secondary cooling cycle is analogous to the installation of a central cooling brine supply for the production complex as a whole. The disadvantage of the apparatus being centrally supplied with coolant, cooling water (from the secondary cycles), and vapor is that processes can only be carried out at specific temperature ranges. An alternative to secondary heating/cooling cycles for individual plants is a central heat supply unit with three temperature ranges. Such a concept is shown in Fig. 3.36.

This system shares the advantage of decentralized secondary heating/cooling cycles in that a broad temperature range can be regulated continuously and that heating and cooling is possible at each temperature. It does, however, differ from the decentralized type in that cooling water can not be brought in directly in cases of emergency.

For all energies – cooling brine, cooling water, vapor, compressed air – lines are available in the pipeline system of multiproduct industrial complexes, so that new apparatus can be supplied quickly at any point.

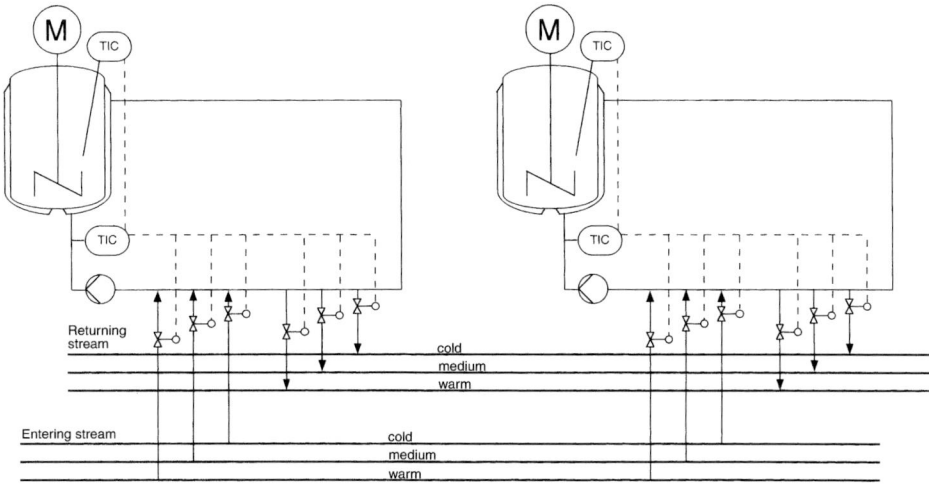

Fig. 3.36 Central heat supply unit with three temperature ranges, according to reference [3.27]

Loss of pressure in long pipes makes extensive supply of vacuum difficult. It is, however, possible to have "vacuum ring roads" in parts of a multiproduct plant, for example, in the vicinity of the distillation columns, thereby both reducing the expense of installing several vacuum pumps and increasing the flexibility in the availability of vacuum.

3.7.4
Peripheral Facilities for Logistics

In some cases, logistical tasks determine a large part of the running of a multiproduct plant. In, for example, the dye-, lubricant-, and also animal-feed-supplement industries, numerous single components are made up into several mixtures and are variously packaged for the market (see also Section 2.3). Peripheral facilities for mixing, packaging, and storage are indispensable for multiproduct plants in these areas. These tasks can even make up the basic marginal operating conditions when such multiproduct plants are planned [3.28]. The tank farm will be dealt with first. Multiproduct plants characteristically have more, but smaller storage tanks compared to those of monoplants. In case regulations on the co-storage of substances need to be complied with, there may be different sections. For greater flexibility, the vessels can often be used not only for products, but also for feedstocks and intermediates. They are connected to the plant by pipeline manifolds.

Stores for solids are built analogously to the storage tanks. The large number of substances and containers necessitate the use of computerized systems for managing the stores. It makes sense to connect these to production planning systems, if these are on hand [3.29].

Mixing and filling stations are further components of the peripheral facilities of multiproduct plants. Here the active substances (e.g., vitamins, pesticide agents, or dye pigments) are combined with auxiliaries, aids, and solvents into products for the market, are packaged according to customer requirements, and are prepared for dispatch. In the agrochemicals field there are plants whose only function is the latter, that is, mixing (formulation) and packaging. For other logistical issues, see also Chapter 9.

3.7.5
Peripheral Facilities for Cleaning

The aim of cleaning between two different product campaigns is to avoid contamination of the following product (cross-contamination). Suitable automated cleaning processes are available, especially for multiproduct plants with their frequent product changeovers. "Cleaning in place" (CIP) (see Section 4.3) is an example of such a rational cleaning procedure; it allows efficient, repeatable, and consistent cleaning of the plant while it is still in production-ready state, that is, the various equipment do not need to be disassembled. The requirement for this is a CIP-oriented plant design, as described in more detail in Sections 4.2 and 4.3, and a CIP supply system as part of the periphery of the multiproduct plant.

The cleaning solutions for cleaning of the production units are circulated through the plant by pumping. They are made available for the sequential cleaning and rinsing steps by the CIP system, which needs to meet the specific requirements of the production plant. There are, in principle, two different concepts [3.30]:

- CIP recovery (multiple use), in which the available cleaning solutions are used more than once
- Single-use cleaning, in which the freshly prepared cleaning solutions are used only once

The CIP recovery unit is usually used as a central supply unit, which cleans different production areas with the aid of different cycles (see Fig. 3.37). It consists of vessels for alkaline and acidic cleaning solutions as well as for fresh and recovered water. Tanks for concentrates, pumps, and valves for the entering and exiting streams of the cleaning cycles and lines with a pipeline manifold also make up part of such a unit. The buffer tanks are usually equipped with integrated heat exchangers, so that the cleaning agents have the required temperature at the start of a rinsing step.

The advantages of CIP recovery are:
- Lower consumption of water, heating energy, and cleaning agents due to the multiple use of cleaning solutions
- Shorter cleaning times owing to the cleaning solutions being kept ready for use

The disadvantages of CIP recovery are the risk of cross-contamination during the cleaning of different active ingredient residues (e.g., in the pharmaceutical industry) as well as possible microbiological contamination of the recovered liquids [3.31]. Validation of such a procedure can therefore be very time-consuming and

Fig. 3.37 Central CIP recovery unit with pipeline manifold and matrix valve arrangement. Reproduced with kind permission of Tuchenhagen

expensive. The concept of "validation" is generally used for the systematic and documented evidence that a specific process or step in a process will consistently produce a product meeting its predetermined specifications and quality attributes.

During single-use cleaning, fresh cleaning solutions are available for each cleaning step. These are pumped through the plant to be cleaned for a specific amount of time and are then treated as waste water to be disposed of or treated. Single-use cleaning units are best placed as preassembled ready-for-use compact units (modules) as close as possible to the area of the plant that needs to be cleaned. They consist of one or two buffer tanks, a supply pump, a dosing pump, valves, dosing and heating installations, and the appropriate measuring and regulating instruments with operating panel, complete with connected tubing and wiring and mounted on a frame (Fig. 3.38). For assembly on location, only connecting to the entering and exiting streams of the cleaning solutions and the energy supply (water, steam, condensate, electricity) should be necessary.

The advantages of single-use cleaning are:
- Cross-contamination is avoided
- By program control, the cleaning solutions are suited to the cleaning task at hand
- The cleaning unit is low-priced and compact.

Fig. 3.38 Single-use cleaning unit. Reproduced with kind permission of Tuchenhagen

3.8
References

[3.1] POLLAK, P., in *Kirk-Othmer Encyclopedia of Chemical Technology, Vol. 10, 4th ed.*, Wiley, New York, **1994**.

[3.2] STEWART, J.C. *Chem. Eng.* **1996**, *2 (Jan)*, 72–79.

[3.3] DOHM, K.-D., THIER, B. *3R International 18(7)*, 480–485.

[3.4] SCHUCH, G., KÖNIG, J. *Chem.-Ing.-Tech.* **1992**, *64(7)*, 587–593.

[3.5] UHLIG, R.J. *Automatisierungstechnische Pra.* **1987**, *1*, 17–23.

[3.6] BRUIJN, W.C., ANTHEUNISSE, A. *PT-Processtech.* **1989**, *44(11)*, 38–41.

[3.7] BESSLING, B., CIPRIAN, J., POLT, A., WELKER, R. *Chem.-Ing.-Tech.* **1995**, *67(2)*, 160–165.

[3.8] CIPRIAN, J., POLT, A. *Chem. Technik* **1996**, *48(6)*, 293–300.

[3.9] BESSLING, B., POLT, A., WELKER, R., *Methodische Ansätze zu Verfahrensgestaltung und Prozeßführung in Mehrproduktanlagen.* GVC Fachausschuß „Prozeß- und Anlagentechnik", Lüneburg, 25–26 October **1993**.

[3.10] BESSLING, B., CIPRIAN, J., POLT, A., WELKER, R., *Verfahrensgestaltung und Prozeßführung in Mehrproduktanlagen.* GVC Jahrestreffen, Aachen, 28–30 September **1994**.

[3.11] WESTERTERP, K.R., VAN GELDER, K.B., JANSSEN, H.J., OYEVAAR, M.H. *Chem. Eng. Sci.* **1988**, *43(8)*, 2229–2236.

[3.12] McGREAVY, C., DE Q. NOVAIS, A.O.L., *Iterative approach to the study of optimal production policies for a multipurpose plant.* CACE Symposium, Montreaux, 8–11 November **1979**.

[3.13] CARLSON, E.C., FELDER, R.M. *Comput. Chem. Eng.* **1992**, *16(7)*, 707–718.

[3.14] PINTO, J.M., GROSSMAN, I.E. *Comput. Chem. Eng.* **1994**, *18(9)*, 797–816.

[3.15] MURAKI, M., KATAOKA, K., HAYAKAWA, T. *Chem. Eng. Sci.* **1986**, *41(7)*, 1843–1851.

[3.16] Unknown *Chemie-Produktion* June **1994**, pp. 22–26.

[3.17] WHITTAKER, R. *Chem. Eng.* 28 May **1984**, pp. 81–88.

[3.18] RHANDHAVA, R., LO, R.N. *CEP* **1982**, *(Nov)*, 76–81.

[3.19] GAILLIOT, F. P., FUTRAN, M., SIGAL, C. T., WILSON, J. J. *Chem. Eng. Prog.* **1987**, *10*, 55–58.

[3.20] Anonymous *Process* **1995**, *7(8)*, 56.

[3.21] NIWA, T. *Chem. Eng.* **1993**, *100(6)*, 102–108.

[3.22] HASEBE, S., Lecture at BASF AG, Ludwigshafen, 25 October **1993**.

[3.23] FÜRER, S., RAUCH, J., SANDEN, F. *Chem.-Ing.-Tech.* **1996**, *68*, 375–381.

[3.24] NOELTNER, G. *Chem.-Ing.-Tech.* **1982**, *54(12)*, 1087–1091.

[3.25] ARIMA, M. *Kgaku Sochi* **1986**, *28(11)*, 43–49.

[3.26] GERASIMOV, V. S., BALAKIN, I. M. *Khim. Neft. Mashinostr.* **1993**, *3*, 9–13.

[3.27] THIER, B. *3R International 24(11)*, 648–657.

[3.28] BARGMANN, J., personal communication.

[3.29] Anonymous *Verfahrenstechnik* **1996**, *30(5)*, 94–95.

[3.30] HIELSCHER, C. *Fat. Sci. Technol.* **1989**, *91*, offprint.

[3.31] FRIEDRICH, G. *Pharma-Technologie-Journal* **1994**, *15(1)*, article no 1069.

Part 2
Planning and Operating Multiproduct Plants

4
Machinery and Apparatus

4.1
Introduction

The function of a piece of apparatus is to enclose the space in which a process takes place, to demarcate the process space from the surrounding area, and to safeguard the process technological and operating conditions and the procedures of the relevant processes. The elements making up the apparatus are determined by their static capacity, the extent of the influence of energy and material transport, as well as the connections between the individual apparatus to process units directly preceding and following them [4.1].

The trend towards standardization and the use of modular systems for the most commonly used apparatus and machinery (e.g., agitated vessels, columns, heat exchangers, pumps) has, among others, been influenced by the need for rational construction of apparatus for process engineering and for simplified storage management. At the same time, continuous improvement in the relevant process functions, such as heat and material transport, the specific energy requirements (power consumption), the separation-, mixing-, or throughput capacity, is strived for.

Whereas these factors are basically valid for the entire apparatus technology of the chemical industry, that is, for both conventional monoplants (dedicated plants) and multiproduct plants, the apparatus found in multiproduct plants have additional specific characteristics, influenced by various factors, as shown in Fig. 4.1. The specific characteristics result, on the one hand, from the specific requirements with regard to the functions of the different plant concepts (see Chapter 3); the plant concept, in its turn, is determined by the pertinent application areas (research and development, production, or product classes, see Chapter 2), as well as the appropriate application fields (operative ranges) (see Chapter 3). On the other hand, the specific characteristics result from the specific basic requirement placed on multiproduct plants by the different material properties of the different starting materials and products (e.g., density, viscosity, suitability for filtration, toxicity, corrosiveness), the different operating conditions of the various processes (e.g., temperature, pressure, pH, volume, volume flow), the current market specifications (purity of products, availability of products, e.g., "just in time"), and the de-

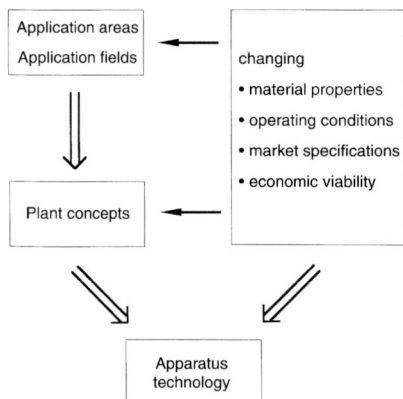

Fig. 4.1 Influences on the specific characteristics of the apparatus technology of multiproduct plants

sired economic efficiency of the plant (e.g., minimal costs associated with product changeover).

Material properties, operating conditions, market specifications, and economic considerations also influence the application areas and fields for multiproduct plants and the choice of a suitable plant concept (see also Chapter 11).

4.2
Basic Requirements and Constructive Solutions

Whereas the apparatus technology of the various multiproduct plant concepts differ in some construction aspects, the following basic requirements are common to all plant concepts, resulting in common construction solutions:
1. The extensive ability to deal with substances with widely different physical and chemical properties (i.e., product-assortment flexibility)
2. The extensive ability to deal with different processes and operating conditions (i.e., product-assortment flexibility and flexibility in capacity)
3. Minimal time consumption and expense resulting from product changeovers
4. Constantly high and consistent product quality.

Adaptability to Products and Processes
An extensive ability in the first two of the above requirements, namely, adaptability to products, processes, and operating conditions, can be reached if equipment such as agitators and agitator drives, pumps, and motors are able to handle a wide range of densities and viscosities, or at least deal with the relevant conditions. The application-oriented *performance adaptations* required for this are brought about by frequency-regulated electric motors; frequency regulation makes it possible to adapt the number of revolutions to the agitator and the mixing function or the pump characteristics. Specific automation functions, such as constant delivery pressure, or constant flow rate, can be fulfilled with appropriate auto-

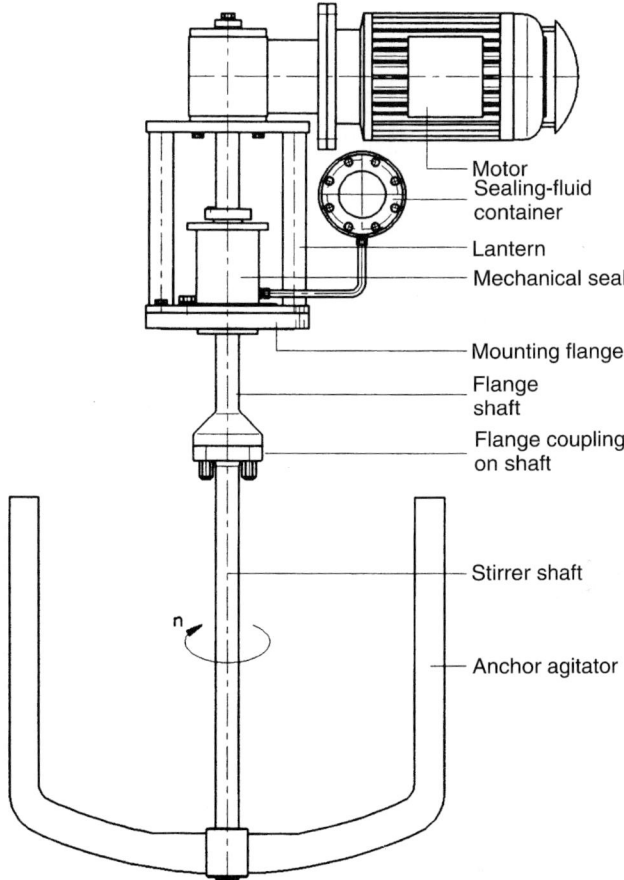

Motor
Sealing-fluid
container

Lantern
Mechanical seal

Mounting flange

Flange
shaft

Flange coupling
on shaft

Stirrer shaft

Anchor agitator

Fig. 4.2 Agitator assembly with flange coupling on the shaft for exchanging the agitator. Reproduced with kind permission of Fluid

matic controls. For the various unit operations that can be carried out in an agitated reactor vessel, there are different suitable agitators available: for example, flat-blade disk turbines for reactions involving exposure to gas and for dispersing; pitched blade turbines or propeller agitators with stream breakers for blending and suspending; and helical ribbon impellers or anchor agitators for good heat transfer to the walls of the reactor vessel. Agitators are therefore equipped with flange couplings on the shaft (see Fig. 4.2), so that it is relatively easy to exchange the agitators. For optimal mixing at different volumes (flexibility in capacity), two- or three-stage agitators are equipped with sliding hubs on the shaft, for adjusting the height of the blade turbines (see Fig. 4.3).

How the construction of pumps can be modified to deal with the flexibility requirements related to the transport of substances is described in Section 4.5.

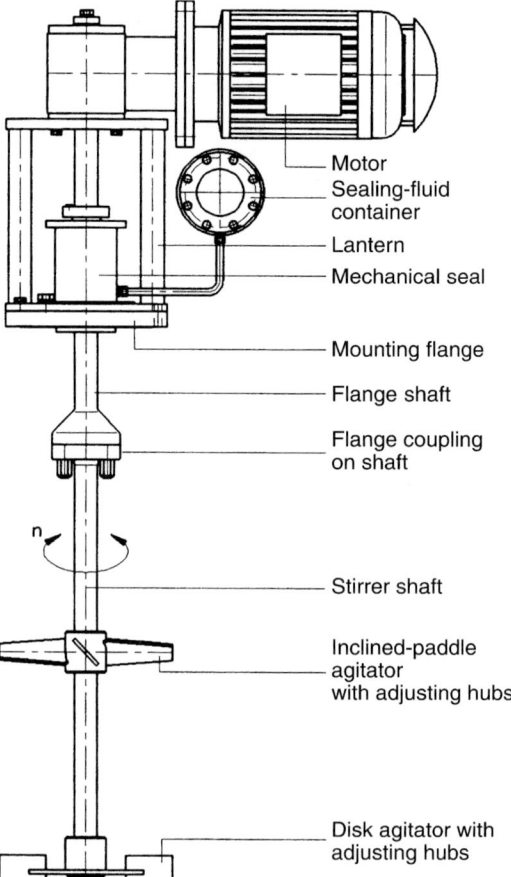

Fig. 4.3 Agitators with sliding hubs for adjusting the height of the blades. Reproduced with kind permission of Fluid

Motor
Sealing-fluid container
Lantern
Mechanical seal

Mounting flange
Flange shaft
Flange coupling on shaft

Stirrer shaft

Inclined-paddle agitator with adjusting hubs

Disk agitator with adjusting hubs

The requirement that the technology of the machinery and apparatus should be suitable for different products and processes also places special demands on the *tightness* of the equipment: for example, a connection that is opened and closed between products, for cleaning or for constructing a new plant functionality for the next process, should still close tightly after being opened and closed several times (see also Section 5.3, Flanges, Couplings, and Seals). The sealing material is also exposed to various substances and temperatures and should therefore be as universally resistant as possible. After all, the simultaneous handling of flammable, highly active, or toxic substances in the various equipment of a multiproduct plant presupposes a very leakproof operation. Just as contamination of the environment and the operating personnel should be avoided, the opposite also applies: for example, for processes operating under vacuum, contamination of the products by the environment, such as by foreign substances, products, microbes, oxygen, or other non-inert and undesired components of the surrounding air, should also be prevented. To meet these requirements, multiproduct plants need to be constructed as closed systems. Where this is not possible, an alternative solution is the construction in compartments or rooms with separate air supply and exhaust.

Avoiding seals altogether is the best way of ensuring leakproof operation. This means that joints of metallic vessels, pipelines, and valves should be welded shut as far as possible. In agitated vessels, the possibility of closing up the torispherical dome by welding it shut, should be considered, to avoid the need for a main flange and seal.

With nonmetallic pipelines (e.g., glass) or lined pipelines (glass-lined steel, steel/PTFE, etc.), special parts are necessary, especially for T-pieces, inlet distributors, valve and fitting connecting pieces, and their combinations; these are increasingly becoming commercially available.

With pumps, the sealless canned-motor pumps and magnetic-drive pumps are preferable to pumps with mechanical seals. Their use in the case of solid-containing media and where cleanability is required is, however, limited because of their very narrow openings. For this reason, special double-acting mechanical seals with pressurized sealing liquid are used in pumps, and also in the drives of agitators. The applied pressure reliably prevents, especially in the case of additional monitoring, a leak to the atmosphere. If an inert sealing liquid (e.g., white mineral oil, completely deionized water, or even completely purified water under nitrogen) is used, the minor loss of sealing liquid in the production room usually presents no problem.

Where the use of seals cannot be avoided, seals under static strain are preferable to those under dynamic strain. For example, with valves, a corrugated-bellows-type construction is better than a packing box. Seals in pipelines, the demands placed on these, and examples in the practice are dealt with in greater detail in Section 5.3.

The avoidance of seals, as far as practicable, not only results in the prevention of leaks, it also allows the requirement of *cleanability* of multiproduct plants to be met (discussed further on). This is because every seal, with its inevitable gap, presents a potential dead space that cannot be completely drained or rinsed.

Because the equipment is subjected to an extensive range of products and processes, the *materials* used in the technical equipment need to be exceptionally corrosion-resistant. All the materials used should have good chemical, mechanical, and thermal resistance towards the products as well as the cleaning solutions. For the metallic materials, chrome nickel steel of the following quality is mainly used:

| Material number according to: | | |
DIN (German industrial standards)	AISI (American Iron and Steel Institute)	
1.4401	316	Standard stainless steels
1.4571	316 Ti	
1.4404	316 L	Low-carbon steel
1.4435	316 L	

Specialized steels such as Hastelloy

The only difference between the compositions of the DIN 1.4401 and 1.4571 materials is that the latter contains titanium, which has a stabilizing effect on the crystal structure, which improves the corrosion resistance, important mainly along the welding seams. In the pharmaceutical sector, there is a trend away from the traditional materials (DIN 1.4401 and 1.4571) to the DIN 1.4404 and 1.4453 materials. The low carbon content of these low-carbon steels imparts the material with good corrosion resistance upon welding, and is good for materials with electropolished surfaces. The DIN 1.4571 stainless steel is, in contrast, not electropolishable. For reasons related to cleaning technology, but also because of the dangers associated with biocorrosion, rolled steel materials with sealed, uncharged, and less reactive surfaces, and those not abrasion- and cavitation-sensitive (pumps) are increasingly replacing cast steel [4.2].

For parts that do not come into contact with products, such as casings, jackets, motors, bearings, bolts, and so forth, the DIN 1.4301 material (AISI 304) is usually acceptable.

Because of their high corrosion resistance, especially towards acids, but at low temperatures also against alkalis, glass-lined steel and glass are also widely used. Their application is basically restricted by their maximum thermal-shock resistance and also, in the case of glass, by the maximum allowable operating overpressure (usually 1 bar). These materials do, however, have the advantage over similar resistant special materials, such as Hastelloy steel, titanium, and tantalum, that they are less expensive.

Materials suitable for use in multiproduct plants as well as nonmetallic lining and coating materials will be discussed in more detail in Chapter 6.

A further basic requirement for the apparatus used in multiproduct plants is that the temperature and pressure ranges covered should be as broad as possible. The appropriate stability design should therefore be incorporated in the planning of the multiproduct plant. The *design and the dimensions of the building parts* should be consistently chosen to fulfill all the requirements set by the operating conditions of all the processes and products of that particular plant.

For reaching the required operating temperature, the apparatus should be equipped with double shells or half-pipe coil jackets for indirect heating or cooling. It should also be possible to control the temperature of pipelines carrying products (especially forced circulation cycles) and valves and fittings; for this purpose they should have double shells or accompanying heating.

Adaptability to Economic and Market Demands

Multiproduct plants meeting requirements 3 and 4 listed above need to produce economically and be adaptable to market requirements. A basic prerequisite for the economic running of a multiproduct plant is that the time and expense resulting from product changeovers should be minimal. For this it is essential that, firstly, the work involved in adapting or changing over the plant should be minimal and, secondly, that the cleaning effort should be minimal. The first of these is achieved by the choice of a suitable plant concept, based on the application area, the product groups, and the application fields and limitations (see also Chapter 11).

The basic requirement of minimal cleaning effort, that is, optimal *cleanability* of the multiproduct plant, has several implications with regard to the apparatus technology, and technical concepts and constructive solutions valid for all plant concepts follow from this.

Fig. 4.4 elucidates the route from economic and market demands, via basic requirements, to the technical concepts. If the basic requirement of optimal cleanability of a multiproduct plant is met, the time required for cleaning between different products, and thus the plant's downtime and costs are kept to a minimum. The cleanability of a plant is optimal if the plant can be completely drained between products. This also ensures economic running, as product losses, cleaning-agent consumption, reconditioning and disposal costs, waste-water contamination, and the need for storage are kept to a minimum.

The complete draining and cleaning of a plant can also prevent cross-contamination of products, thereby ensuring high and consistent product quality (market requirement). [*Cross-contamination* is the contamination, after production has been switched from one product to another (product changeover), of the new product by residues of the previous product.]

The basic requirement that a plant should be capable of being completely drained and therefore optimally cleaned is best met by apparatus, machinery, pipes, and valves whose technical design ensures complete absence of dead space as well as smooth and clean surfaces. A less suitable alternative is a plant that is easily assembled and dissembled – here complete draining and thorough cleaning is also possible, but product loss and more time and cost expenditure are unavoidable.

Avoiding contamination of the product, especially cross-contamination, is one of the basic requirements of the GMP guidelines for the manufacture of pharmaceu-

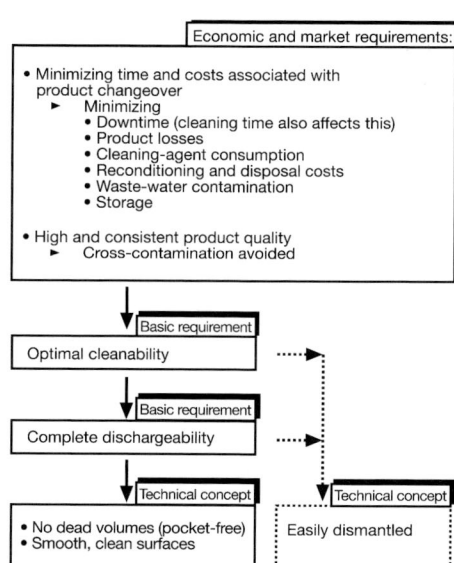

Fig. 4.4 The route from economic and market demands to technical concepts

tical agents and medical products. GMP operation means the consistent and safe execution of processes by procedures with controlled parameters, in a defined, constant environment. Products with high and consistent quality result from this. Multiproduct plants producing high-value products all have very similar requirements, especially in the pharmaceutical, foodstuffs, and fine chemicals sectors, and therefore the apparatus used in these plants are increasingly constructed with GMP characteristics with regard to the absence of dead space and easy cleaning.

Standard solutions in construction to meet these requirements are not widely available yet; the manufacturers of apparatus, machinery, valves, and pipelines usually, often in conjunction with the client, develop new constructions or transfer known construction principles to new applications. An example of this is the design of flange joints and seals: the flanges of the vessels of metallic apparatus can be made as dead-space-free as pipeline flanges (see Section 5.3), through a construction with an O-ring seal in a groove around the edges (Fig. 4.5). A maximum defined elastic deformation by means of a metallic stop will ensure that the O-ring is long-lasting. In contrast to the standard construction used in chemical apparatus so far (Fig. 4.6), where a flat seal is placed between the flanges with groove and tongue, a circular gap is avoided. Such gaps always attract product residues, which are often not even removed by extensive cleaning procedures with high-pressure cleaning nozzles directed through the manhole. The gap-free O-ring seal around the edges allows, in contrast, cleaning of the closed apparatus (i.e., CIP, cleaning in place).

Dead volumes can also be avoided and cleaning can be made easier if all the ports (sockets) for the agitator assembly, safety equipment, manometers, inlet connections, and so forth, are kept as short as possible, and instead of going through (see Fig. 4.6), are attached to the equipment (Fig. 4.7). Where sterilization with steam is required (e.g., pharmaceutical industry), it should be noted that the required sterilization temperature may not be reached in a port (socket) whose length is three times greater than its diameter.

Outlets of vessels should, in principle, be as close to the base as possible, without dead volume, and be easy to clean. Very different constructions are available for this purpose, and their advantages and disadvantages will depend on the specific application. Ball valves and diaphragm valves in block-flange form seal off

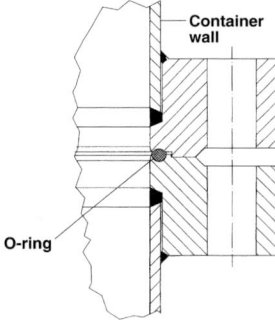

Container wall

O-ring

Fig. 4.5 A vessel with a flange with an O-ring seal. Reproduced with kind permission of ASCA-Metallwarenfabrik

Fig. 4.6 Standard construction of chemical apparatus with flanges with groove and tongue and flat seal

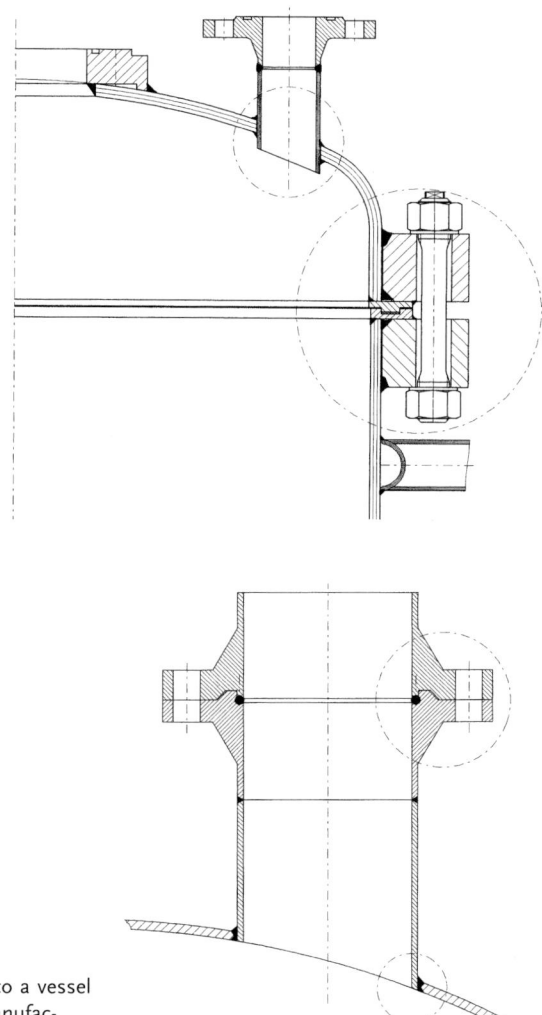

Fig. 4.7 A dead-volume-free socket to a vessel head with an aseptic flange joint, manufactured by Guth

close to the bottom. The body of the valve is sealed against the block flange of the vessel, for example, with an embedded O-ring, but can also be directly welded onto the base of the vessel, doing away with the need for this seal, with its dead volume. The advantage of the ball valve, namely unhindered throughput, may also be a disadvantage: even though so-called pocket-free ball valves supposedly have no dead space in the closed state, as no product is contained in the ball bore, they are not hundred percent dead-volume-free; through rotation of the ball, product can work its way between the shell of the bearings and seal (made of PTFE) and the external ball surface. This dead volume can only be reached by CIP or SIP (SIP, "sterilization in place") in special cleaning positions of the ball.

The product space of the diaphragm valve is, in contrast to that of the ball valve, hermetically sealed against the surroundings by the fixed diaphragm. This ensures absolutely aseptic operation, as required in the food and pharmaceutical sectors. The use of diaphragm valves is, however, severely limited by the wear-and-tear of the rubber-elastic membrane (PTFE-coated on the product side) in case of high operating temperatures and large nominal diameters of the valve.

Disk valves (see Fig. 4.8) are well-proven fittings for vessel outlets. Their considerable advantage is that the seal of the seat of the valve is wear-resistant. The casing can also here be welded directly into the torispherical base (in exclusively metallic vessels) and seal right to the base. Hermetic separation between product space and surroundings, resulting in aseptic operation, is also made possible by corrugated bellows (e.g., of PTFE or stainless steel). The sealing part of the valve may also be equipped with a temperature sensor (see Fig. 4.8). This site has the advantage that even in relative empty state, for example, when a stream breaker with temperature sensor no longer reaches the liquid, reliable temperature indication is still possible (flexibility in capacity).

Important for agitator assemblies is the design of the seal of the shaft. Packing boxes are a thing of the past; well-proven since are double-acting mechanical shaft

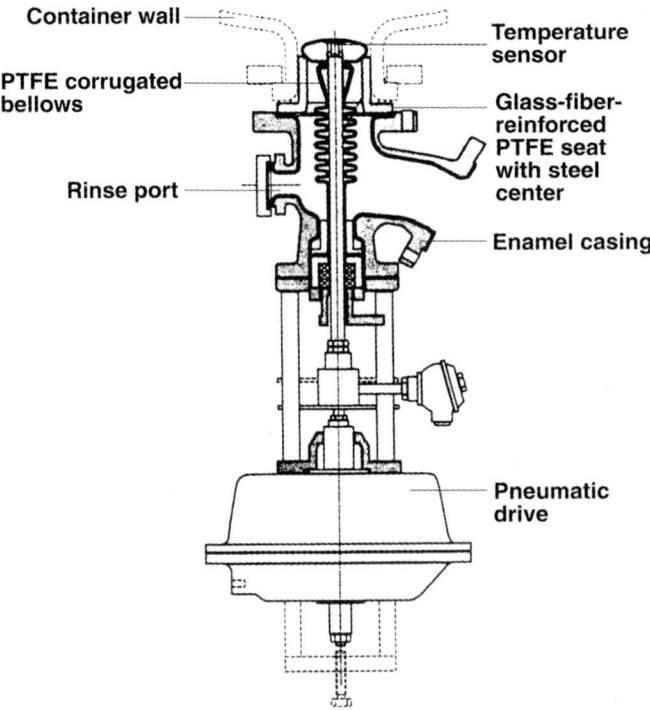

Fig. 4.8 A valve placed at the base, with corrugated bellows and temperature sensor in the seal section (De Dietrich)

seals. The lubricant necessary for the seal ring acts at the same time as barrier between product and atmosphere. As the possibility of minor leakage into the product space cannot be completely excluded, a medium that does not contaminate the product should be chosen.

Mechanical shaft seals have several advantages. The agitated vessel can be stirred under pressure and at higher temperatures if the barrier medium (sealing liquid) is pressurized and cooled (e. g., by convection to thermosiphon container). Mechanical seals are adaptable with regard to the material used for the seal rings and the choice of lubricant. Mechanical seals can be sterilized.

Special constructions with regard to details are seen in the designs of agitators. For example, the flange couplings on the shaft and the sliding hubs (as described much earlier, see also Fig. 4.3) may be constructed completely dead-volume-free and CIP-cleanable through suitable geometries and O-ring seals.

Avoiding dead space is also an important requirement for measuring instruments and their installation, so that cleaning is facilitated. The use of in-line technology, as already applied in the food and pharmaceutical industries (Fig. 4.9), has removed low-flow zones in T-pieces and connecting pieces for measuring devices. The in-line casings are welded into metallic piping or are connected by flange joints (see Section 5.3) with little dead volume. Apart from the suitability of these systems for CIP/SIP, the standardization of the connections to the devices is a further advantage. In-line technology is available for measuring pressure, temperature, conductivity, flow, turbidity, and color (color switches), and the installation of sight glasses and sampling devices is also possible.

Suppliers of glass-lined steel and glass apparatus and piping are increasingly producing special constructions that contain no narrow gaps at the seals, flanges,

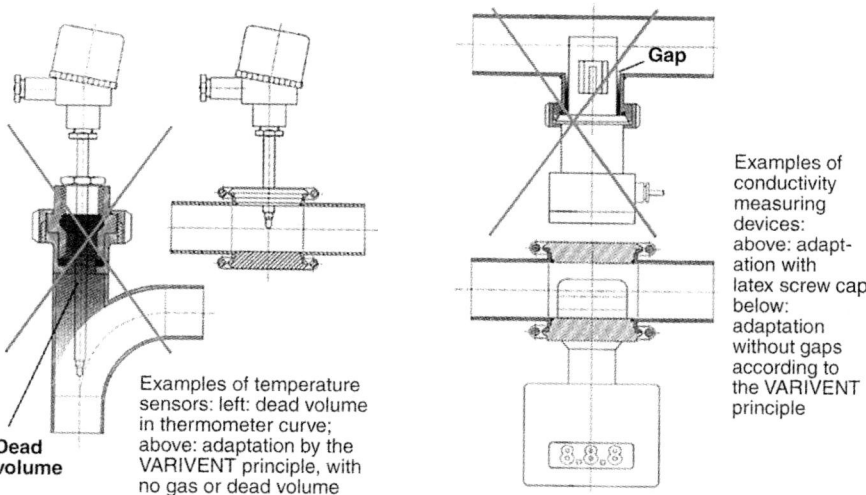

Fig. 4.9 Dead volume is avoided by the use of in-line measuring instruments. Reproduced with kind permission of Tuchenhagen

Fig. 4.10 Horizontal skimmer centrifuge for pharmaceutical applications. Reproduced with kind permission of Krauss Maffei

and vessel sockets, can be easily emptied at suitable gradients, and contain no dead volumes that cannot be easily rinsed out.

The same principles are valid for machinery, but their application is limited if they result in restrictions in the functioning of the machine. In such cases, large inspection openings and/or the disassembly of built-in parts and movable machine parts (e.g., with centrifuges, Fig. 4.10, mixers, driers, etc.) allow complete cleaning. Fig. 4.11 shows, as a further example, a rotary gate valve that is easily disassembled and cleaned, also when fitted. Covers, manholes, and inspection openings are usual-

Fig. 4.11 Rotary gate valve. Reproduced with kind permission of Gericke

ly equipped with quick-operating seals, so that the production personnel can carry out cleaning and simple maintenance without the help of craftsmen.

4.3
Cleaning in Place (CIP)

Cleaning in place (CIP) means the automatic cleaning of a production unit on location and in production-ready state, that is, without special handling or disassembly of plant- and equipment parts [4.3].

The precondition is that the design and construction of the plant should, as described in the previous examples, make cleaning easy. The purpose of cleaning between two product campaigns is to avoid contamination of the next product B by residues of product A from the previous run (cross-contamination) or by residual cleaning agents. The cleaning efficiency can be judged according to physical, chemical, and biological criteria.

CIP has already been applied in the food, dairy, brewery, and wine-making sectors for more than 35 years [4.4]. Recently, this cleaning method has also found application in pharmaceutical, biotechnological, and other batch processes. This method has attracted rather less consideration in other areas of the chemical industry, such as in the production of agricultural and petrochemical products. However, the increasing personnel, energy, and auxiliary-agent costs, as well as the more stringent quality requirements are strong economic arguments for the application of CIP in the chemical industry as well. This is especially valid for multiproduct plants producing high-value products.

CIP has several advantages over conventional cleaning procedures such as rinsing, flooding, steaming, steam-jet cleaning, and hand-cleaning. Combined with a suitable plant design, CIP is superior to any other cleaning method: the cleaning result is defined and reproducible and thereby leads to higher production reliability.

Contamination of products, leading to product loss or expensive processing can be avoided by CIP. Since the CIP cleaning process is automated, operating mistakes or inadequate hand-cleaning, for example, is avoided; this also decreases personnel numbers and plant downtimes, which would have been necessary for the dismantling, cleaning, and reassembly of the plant parts. This results in a reduction in operating costs and an increase in the plant capacity.

An additional advantage is that the cleaning agent/solvent quantities necessary for CIP is less than for conventional cleaning.

With its controlled, monitored, documented, and reliably reproducible cleaning result, CIP fulfills the GMP (good manufacturing practices) requirements. These are described in regulation guidelines by the FDA (U.S. Food and Drug Administration) and PIC (Pharmaceutical Inspection Convention) and apply to the pharmaceutical and biotechnological manufacture of pharmaceuticals for the American market, and the member states of PIC.

An important advantage is that CIP improves work safety, since cleaning takes place in a closed production unit and operating personnel do not come into contact with dangerous substances, as with dismantling of the unit and subsequent hand-cleaning.

CIP supply units (see Section 3.7.5, Peripheral Facilities for Cleaning) provide the necessary cleaning solutions. The minimum facilities required for preparing washing liquid consist of a cleaning forerun pump, which supplies the necessary pressure for the required forerun quantities, and a dosing pump, dosing the cleaning agent concentrate.

The cleaning of vessels is different from that of pipes, and high- and low-pressure cleaning also differ from one another. The high-pressure cleaning of the insides of a vessel or a tank through a concentrated full jet with high impact energy is mainly a mechanical cleaning process. For this, spray heads whose cleaning jets describe orbital paths are used. The advantage of this cleaning method is that it is effective for extreme dirt, and it also removes strongly encrusted dirt. The disadvantage is that the spray jet is ineffective in shielded areas behind agitators, parts built into tanks, and inlet- and outlet sockets. In production plants with permanent piping, high-pressure cleaning is therefore limited to the few application areas where the types of impurities make the mechanically operating spray jets necessary.

In low-pressure cleaning, all surfaces that came into contact with the product are completely sprinkled. The cleaning rate depends on the concentration of the cleaning agent, its chemical activity, the temperature, and the flow rate. For cleaning of vessels, a special CIP facility is preferably built permanently into the vessel and kept supplied by CIP valves [4.5].

Fig. 4.12 shows low-pressure spray nozzles available for different spray patterns. A standing vessel with agitator assembly, for example, requires two spray nozzles, one on each side of the agitator, each completely covering the area around itself.

Fig. 4.12 Low-pressure spray nozzles for the cleaning of tanks, vessels, and containers. With kind permission of Tuchenhagen

For larger vessels and tanks, cleaning with a hydraulically operated rotating jet cleaner is a better option (Fig. 4.13): the throughput rate required from these is lower than that required from spray nozzles, because the total spray liquid that emerges from one to three fan-shaped spray jet nozzles spends a fraction of time being directed at only a small segment of the vessel wall. Put differently, for the same throughput rate, the contact spray density with cleaning agent is considerably greater than with a spherical spray nozzle; this, together with the increased turbulence resulting from impulse–pause spraying, leads to shorter cleaning times [4.3].

Thorough cleaning of a vessel always involves the cleaning of pipes. All sockets of a vessel must be integrated into the CIP process. CIP inlet and outlet valves are thus incorporated into the product pipelines, so that no dead cleaning spaces develop. For thorough cleaning of pipes, a defined fluid flow and minimum flow rate ($v > 2\ \mathrm{m\ s^{-1}}$) is necessary, so that the chemical cleaning process is mechanically supported.

The design of a production plant complies with CIP requirements if the following criteria are met:

- All inner surfaces are completely wetted: there are no dead spaces; there are no gaps at the seals, flanges, and sockets; the vessels have no areas shielded from the cleaning sprays; there are no air bubbles in the pumps, the heat exchangers, the pipelines, the valves, the sockets of the apparatus, and the seals
- The unit can be completely drained: if required, the outlets are at the base of the apparatus at the lowest point; the pipes run at a gradient; the number of flanges and couplings is kept to a minimum (as far as possible); these are preferably replaced by welded connections instead (preferably automatic welding processes, e.g., orbital welding)

Fig. 4.13 Hydraulically operated rotating jet cleaner. With kind permission of Tuchenhagen

- Corrosion-resistant materials (e.g., glass-lined steel, glass, stainless steels, special materials, PTFE) are used for the parts of the plant unit that come into contact with the product and cleaning agents (surfaces and seals)
- High-quality finishes (rounded edges, medium surface roughness of $R_a < 0.8\ \mu m$) are used for surfaces that come into contact with products and cleaning agents.

4.4
Established Types of Machinery and Apparatus for Multiproduct Plants

4.4.1
General

How does one choose apparatus and machinery that fulfill the flexibility requirements described in Section 1.1 for a multiproduct plant? This question, in its most general form, is certainly a theoretical one; in practice, however, certain flexibility requirements will be associated with other incidental conditions. With the aid of the flow chart in Fig. 3.2 (Section 3.1), a list of apparatus for an equipment-oriented multiproduct plant can be compiled.

A complete process can be made up by combining together processes from this flow chart or parts thereof. In the most flexible type of multiproduct plant, all of these processes should be possible. Equipment-oriented plants of this type are especially useful when the product spectrum to be produced is not well known. The emphasis in such a case is on structural flexibility and product-assortment flexibility. To keep the costs of such a plant within limits, compromises with regard to the type and quantity of the facilities need to be made.

In addition to the strategies presented in Section 4.1 for the optimization of apparatus and machinery, the following principle is especially important here: in multiproduct plants, apparatus that allow the most possible unit operations to be carried out with the least possible apparatus are preferred.

The workhorse of the multiproduct plant, the agitated reactor vessel, will be used here as an example. In such a vessel, the unit operations mixing, temperature control, reactions, distillations, and (even if only simple) extractions and crystallizations are possible (see Section 3.1.2.2). From solid-process technology, agitated pressure nutsch filters have also been introduced. In these apparatus, first of all, solid–liquid separations can be carried out. But with it, repeated resuspending and washing of crystals are also possible. With some designs, drying of the product is also possible, when the apparatus is operated like a shovel drier. Combinations of centrifuges and driers in one machine (combined apparatus) also fulfill the requirements of multiproduct plants [4.6].

4.4.2
Examples of Established Types of Machinery and Apparatus

If one follows the flow of the product from top to bottom through the discontinuously operated multiproduct plant, the facilities described in the following would be encountered. These correspond to the unit operations represented in Fig. 3.2. At the highest level, several loading vessels for storage or preparation of feedstocks or intermediates are found. They come in a logical arrangement of sizes. The volume of the largest should at least correspond to the total content of the agitated reactor vessel.

The materials of which the loading vessels are made are similar to those used for the reaction vessels, although simpler materials are also used where the requirements allow it. For a glass-lined steel agitated vessel, for example, there would certainly be some glass-lined steel loading vessels, but also ones from stainless steel. At least some of the loading vessels would be equipped with circulation pumps. The pumps allow the content of the loading vessels to be transported to other plants of the operation.

There would be at least a reflux condenser mounted on top of the agitated reactor vessel, although this is often supplemented with a short distillation column. The reflux condenser or overhead condenser would have a heat-transfer surface comparable to that of the reactor vessel. Only a few separation plates are possible with these mounted columns. A phase separator is integrated into the return flow pipe of the condenser for cases such as esterification, where the water of the reaction needs to be removed.

A vacuum unit is usually found at the condenser level. The robust steam-jet/liquid-ring vacuum pump combination units have proved themselves under the constantly changing operating conditions and products. Corrosion-resistant versions of these units are also available, where the steam jet and heat-transfer sections are of graphite, and the liquid-ring pump is of Hastelloy steel. Fig. 4.14 shows the basic construction of such a combination vacuum pump.

Because these vacuum pumps pollute the waste water rather severely, even when run in circulation mode, with indirect cooling, they have attracted strong competition from other concepts, such as the rotary slide-valve vacuum pump and the rotary piston pump [4.7]. The casings of rotary slide-valve and rotary piston pumps are, however, usually constructed from ductile cast iron (GGG 40), which rather limits their corrosion resistance. Such units only operate simply and reliably in conjunction with a whole series of monitoring devices and process control. A controllable vacuum of between 25 and 600 mbar can be obtained.

Solid–liquid separations take place below the agitated vessel. For this, an inverting filter centrifuge is usually used. Its suitability extends, for example, from the separation of solids that are difficult to filter, to solids that form a dense bottom layer, as the filter cake is discharged without a residual layer. CIP attachments are integrated in these machines. Fig. 4.15 shows the construction and operation of an inverting filter centrifuge

Solid–liquid separations can also be carried out with agitated pressure nutsch filters (Fig. 4.16), also known as process filters or filter driers. They belong to the

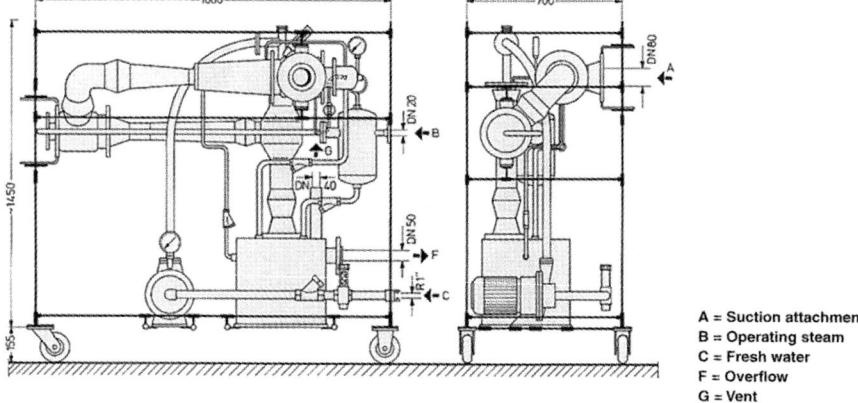

A = Suction attachment
B = Operating steam
C = Fresh water
F = Overflow
G = Vent

Fig. 4.14 The basic construction of a combination vacuum pump. With kind permission of Gea Wiegand

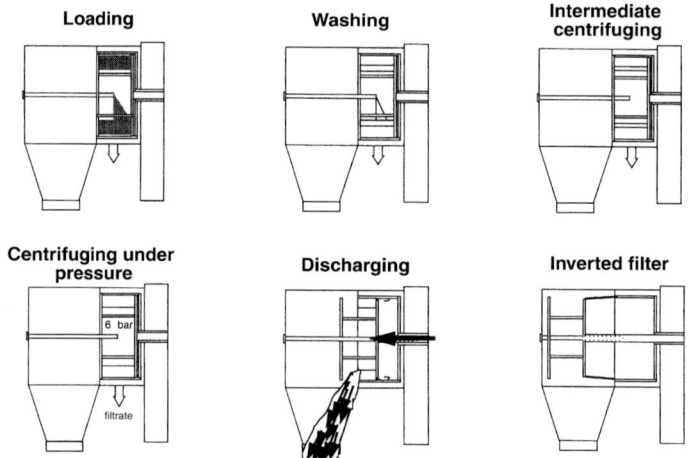

Fig. 4.15 Schematic representation of construction and operation of an inverting filter centrifuge. With kind permission of Heinkel

group of multipurpose apparatus (for more on this, see also Section 3.6). The fundamental part of this apparatus is a pressure nutsch, equipped with a heating and a cooling mantle. These can be rotated about the horizontal axis. A stirrer is also part of the apparatus. The type of filter material used depends on the application, and can be one of various cloth or sintered metal filters. Fig. 4.16 shows how a rotatable pressure nutsch filter is constructed and functions. Various basic operations can be carried out in this apparatus: solid–liquid separations, displacement washing of the filter cake, diffusion washing of the filter cake after resuspension, and, finally, drying. The filter base of process filters can usually be rotated out hydraulically; this is very useful for cleaning.

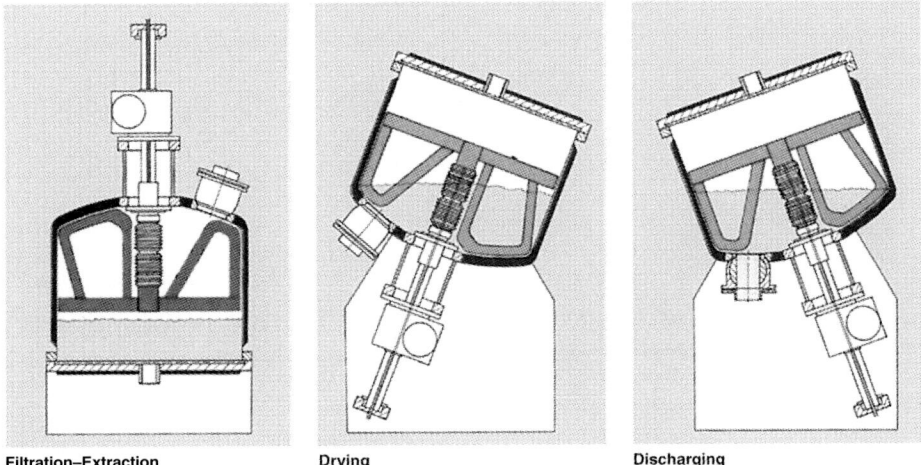

Filtration–Extraction   Drying   Discharging

Fig. 4.16 Basic representation of a rotatable pressure nutsch filter. With kind permission of Schenk

For the drying stage, a dedicated drier can be built in underneath the solid-liquid separation unit. The construction is shown schematically in Figure 4.17. It is necessary for the drier to be as universally applicable as possible. A mixing/drying combination for the drying of bulk goods often makes sense. The construction of such a mixer/drier is derived from that of a shovel drier or, also, a conic drier (conic mixer). As driers often have relatively complicated rotating built-in parts (such as paddles, shovels, etc.), it is important to make sure that they can be drained completely. For flexible control of the drying conditions, heating at var-

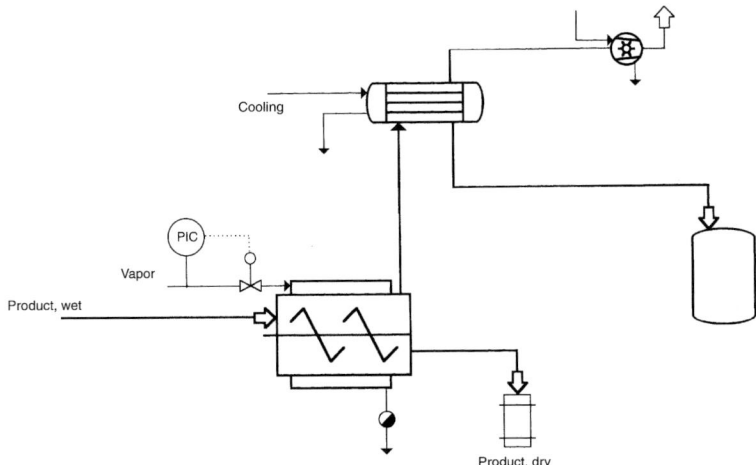

Fig. 4.17 Flow chart of the construction of a drier

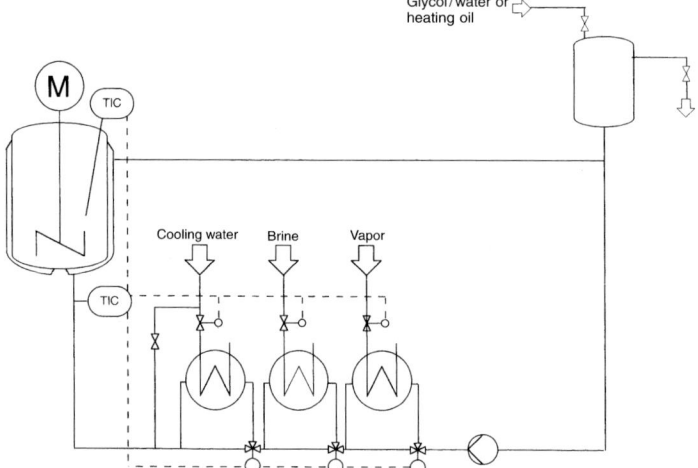

Fig. 4.18 Construction of heating–cooling cycles

ious temperature ranges should be possible. For the same reason, a vacuum unit, similar to that of the agitated vessel, is required. A condenser and receivers for removal of the residual moisture from the material to be dried are also found on the periphery of the drier.

An important auxiliary facility in a multiproduct plant is a secondary heating–cooling cycle. Because flexibility is required, it is especially important that an as wide as possible temperature range in the agitated reactor vessel can be regulated continuously. Heating as well as cooling should be possible at each temperature level. This requirement can be fulfilled if the same heat-conducting medium is used for the whole temperature range. For a typical temperature range of –15 °C to 200 °C, either nonpressurized heat-transfer oils or pressurized water–glycol mixtures can be considered. The three-way valves used in the heat-transfer agent stream and the regulation valves of the cooling and heating media are controlled by split-range regulators.

The basic construction of a secondary heating–cooling cycle, by which the temperature of a plant can be controlled, is shown in Fig. 4.18. The use of a heat-conducting oil avoids the more complicated pressurizing of the heat-transfer medium, but has the disadvantage that heat transfer is worse. The use of water–glycol mixtures not only has the advantage of indirect heat transfer, but also that in case of emergency the heating or cooling medium (vapor, water, or brine) can be transferred directly into the heating-agent cycle. This possibility of emergency cooling, should the circulation pump malfunction, can be an important criterion.

4.4.3
Special Apparatus for Modular Multiproduct Plants

In Section 3.3, the basic construction of modular multiproduct plants was described. The central piece of equipment is usually an up to approximately 1 m³-
sized agitated vessel. The properties of the agitated vessel are described in Section 3.1. The machinery and apparatus belonging to the central agitated reactor
vessel are constructed as a mobile module. The need for mobility necessitates a
very compact construction. Fig. 4.19 shows a mobile pneumatic diaphragm pump
as a module for a multiproduct plant. It is driven by pressurized air, which is
available everywhere in a chemical plant, and can be connected very quickly and
securely via a coupling. Safeguarding the pump against running dry or overheating is not necessary, making the fast, flexible use of this robust type of pump possible anywhere.

Fig. 4.20 shows a commercially available mobile module for the scrubbing of
waste gas. Integrated in the module are a liquid jet scrubber, which draws the gas
to be cleaned in, a scrubber column operated in parallel stream, with packing for
intensifying the mass transfer, a scrubber-liquid container, and a circulation
pump. The electrical supply is accessed via an explosion-proof electrical socket.
The waste-gas stream can be connected via a waste-gas pipe. The scrubber liquid
needs to be topped up from time to time or needs to be supplied and withdrawn
continuously.

The scrubber module used as the example above represents the ready-to-use
modules that are commercially available. There is a trend towards modules in
which the required measuring and regulation facilities, and even control units,
are incorporated. The remaining interfaces of the module are then those to the

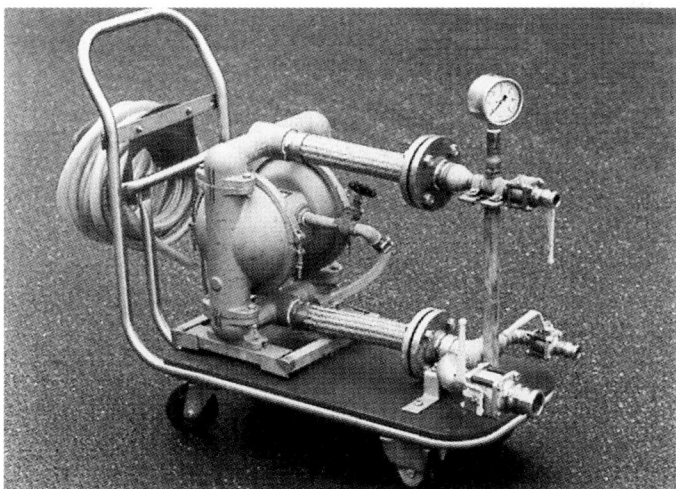

Fig. 4.19 A mobile pneumatic diaphragm pump

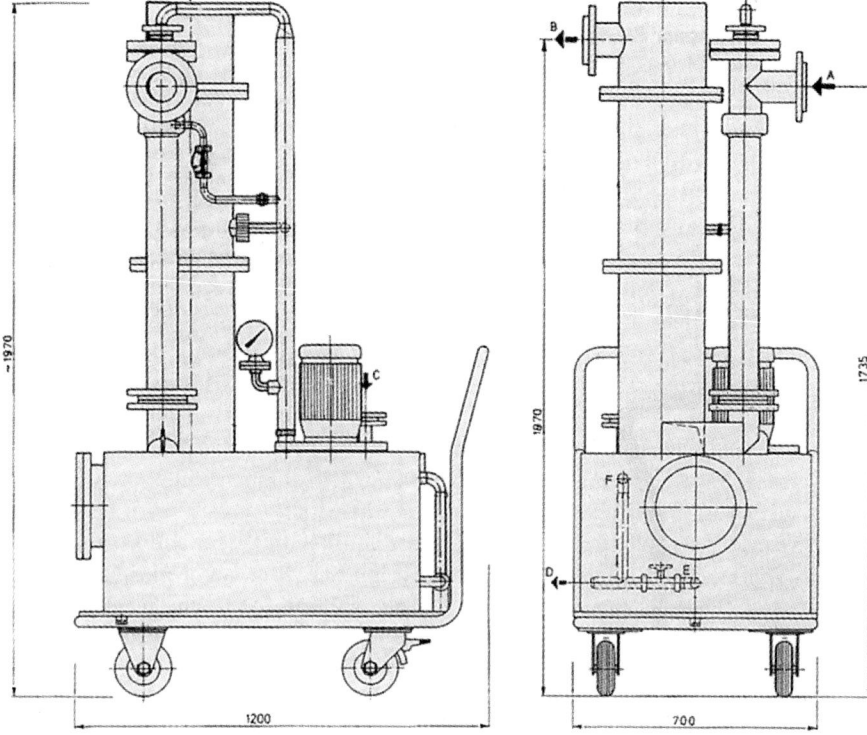

Main measurements in mm

A = Gas entry
B = Gas exit
C = Industrial liquid
D = Outlet

Fig. 4.20 Scrubber module. With kind permission of Gea Wiegand

pipelines (material flow), to the energy supply, and the measurement, control, and regulating interfaces (information interfaces, field bus concepts; see also Chapter 7). All the facilities are combined onto a mobile platform.

If the agitated vessel as central equipment is taken as the starting point, a look at Fig. 3.2 (unit operations in an agitated vessel) shows which of the functions of the main processes in a modular multiproduct plant can be fulfilled by modules. Possible, and also found in practice, are, firstly, dosing modules. Their purpose is to supply the main apparatus, that is, the agitated vessel, with feedstocks in a defined manner. In the case of solids, for example, this can be in the form of a container balance with dosing facilities, or, for liquids, be containers with adjustable dosing pumps or flow meters.

A module generating forced circulation takes care of heating and cooling. Such a module consists of a pump and a compact heat-transfer apparatus. It is simply connected to the vessel, to produce an external heat-transfer cycle, which can, for example, be used for flash evaporation for distillation in the vessel.

Since the process or product determines how the modules are constructed around the central apparatus, and since the modules need to be replaced quickly by other modules at the end of the campaign, thorough cleaning whilst being connected is not necessary. After being disconnected, the module can be dismantled outside the plant, and be cleaned thoroughly where necessary. For the next product, clean modules from the pool are available for the new plant configuration and production. Only the cleaning of the central apparatus needs to be fast and thorough and without further dismantling.

4.5
Pumps

In multiproduct plants, as in all chemical plants, pumps are the machines encountered most often. As with all the other tasks, that of material transport also needs to be highly flexible. The pumps in multiproduct plants need to be flexible with regard to their discharge head, delivery quantities (flexibility in capacity), and above all, with regard to the goods transported (flexibility in product assortment). The pressure at which the pumps operate is usually from vacuum up to approximately 4 bar overpressure on the inlet side plus the discharge pressure of the pump.

Mainly small- to medium-sized pumps are found in multiproduct plants. The limited size of the apparatus and therefore the charge in a multiproduct plant usually already restricts transport quantities to approximately 2–20 $m^3 h^{-1}$. The majority of these pumps have a discharge head of up to 50 m water. Relatively small pumps can also, without problems, run outside the optimal operating range. With a pump installation that includes recirculation into the receiver, considerable flexibility, relative to the nominal specifications of the pump, in quantity and delivery head can be achieved. In Fig. 4.21, such a flexible pump setup is shown.

Adequately monitored canned-motor pumps or magnetically driven (mag-drive) pumps should be used, wherever possible, in multiproduct plants. Monitoring is done to ensure minimum throughput and to protect against running dry. This protecting function is accomplished in modern magnetically coupled pumps, as in canned-motor pumps, through temperature monitoring at the separating can. The narrow slit found in this type of pump generally makes the handling of solid-containing media difficult. Even for solid-containing carrying media and slurries, magnetically coupled pumps are now available; this has in the past been the domain of mechanical seal pumps only. With these magnetically coupled pumps with turned-around magnetic drum, very widely applicable pumps have become available for conveying all media types. These pumps can also be constructed from corrosion-resistant special materials.

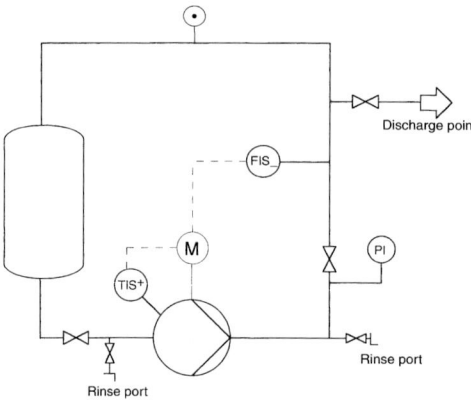

Fig. 4.21 Flexible pump construction in multiproduct plants

The so-called combination pump is a modern solution to the flexibility demands [4.2]. With this type of pump, the casing is permanently installed within the tubing, while the impeller and shaft seal can be adapted to the application. For liquid only, for example, one would choose a closed impeller and a single mechanical seal; for solid-containing media, an open impeller and a tandem mechanical seal with a quench system would be used. With a frequency-controlled drive, the number of revolutions at which the pump runs can be adjusted to suit the requirements.

Another possible strategy for achieving the required flexibility is to use standard models. For this, the course of the tubing should be constructed in such a way that the various pumps can be used for different conveying functions, but fit into the piping in the same manner. Block-constructed pumps are ideal for this, and small to medium-sized ones can even be "hooked onto" the tubing, either on their own foundation, or even completely without an own pump foundation.

Flexibility is additionally increased if the pumps are connected through plugs, also for signal transfer. In such a case, the plant personnel can exchange the pumps as required, without needing to make use of a technical workshop. Cleaning or maintenance work on the pumps can be carried out at a later stage, where necessary. Especially where changing from one product to another takes place, this can result in considerable time gains. Fig. 4.22 shows the foundationless setup of a pump, in blockwise construction, connected via a plug.

The mobile, compressed-air-driven pneumatic diaphragm pump, which can be readily used where required, is also a valuable asset in a multiproduct plant. This pump is self-priming and robust, and when, for example, mounted on a small trolley, can be put into operation fast wherever a pump is needed, simply through a coupling and hose. Fig. 4.19 shows a mobile pneumatic diaphragm pump as a multiproduct-plant module (see also Section 3.3 for further details). If, especially in the pharmaceutical branch, CIP and SIP is required from the pump, pumps with tandem mechanical seals are almost exclusively utilized nowadays. Pressurizing the tandem mechanical seals between the two packing rings reliably prevents

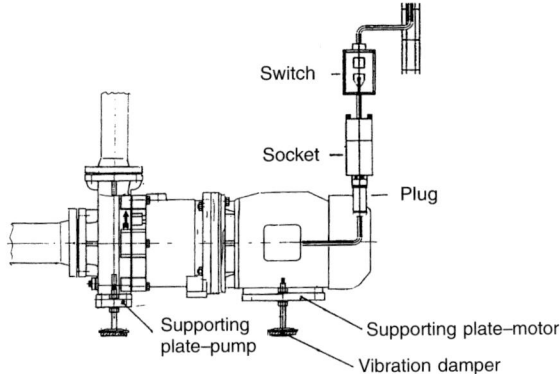

Fig. 4.22 Foundationless blockwise construction of a pump with power supply

(especially where additional monitoring units are present) leakage to the atmosphere. The minor loss of sealing medium into the product chamber is usually not a problem, as long as an inert sealant (e.g., white oil, completely desalinated water, or even purified water under nitrogen) is used. The narrow slits and dead volumes found in canned-motor pumps and magnetically driven pumps make the requirements of easy cleaning and sterilizability difficult to fulfill.

4.6
References

[4.1] VAUCK, W. R. A., MÜLLER, H. A., *Grundoperationen chemischer Verfahrenstechnik*, Wiley-VCH, Weinheim, **1988**.

[4.2] BERDELLE-HILGE, P. *Chem.-Anlagen Verfahren* **1994**, *(Jul)*, 51–52.

[4.3] HIELSCHER, C. *Fat. Sci. Technol.* **1989**, *91*, offprint.

[4.4] STEWART, J. C., SEIBERLING, D. A. *Chem. Eng.* **1996**, *(Jan)*, 72–79.

[4.5] FRIEDRICH, G. *Pharma-Technologie-Journal* **1994**, *15(1)*, article no 1069.

[4.6] GEHRMANN, D., HENSMANN, K.-H., UHLEMANN, H. *Chem.-Ing.-Tech.* **1994**, *66*, 1616–1619.

[4.7] MÜLLER, K. P. *Vakuum in der Praxis* **1994**, *2*, 109–112.

5
Pipelines and Connections Technology

5.1
General

In multiproduct plants, as in all classical chemical plants, pipes are responsible for the transport of materials between apparatus and machinery. Through pipes, materials are conveyed all the way, from one process to another, from one part of the plant to another. However, in multiproduct plants, the routes between the process steps, from apparatus to apparatus, from machinery to machinery, are subject to changes in two ways. Firstly, the order in which the processes are run differ from product to product, and, secondly, the material to be transported differs from case to case. The various pipes connecting the individual operation units are therefore important pillars upon which both structural flexibility as well as product-assortment flexibility rest.

A large portion of the flexibility of conventional multiproduct plants (excluding pipeless plants) is achieved through the design of the pipelines. In multiproduct plants, which are constantly being altered, and therefore have constantly changing connections, the necessary overview is ensured only if the pipelines are planned in such a way that they are clearly laid out and distinguishable. In multiproduct plants, sufficient space in the form of pipeline lanes should be reserved, not despite, but because of the frequent changes and rebuilding.

A central problem associated with pipelines in multiproduct plants is that of cleaning between production campaigns (see also Section 5.5).

Apart from the product routes used by the processes currently run in the plant, the pipeline system in a multiproduct plant usually has further possible product routes available for further processes. This leads to another possible problem, that of safeguarding against the wrong route (see Section 5.6).

5.2
Design of Pipelines and Hose Lines

5.2.1
Pipelines

When planning the piping within a multiproduct plant, the designing engineer faces a task less narrowly defined than would be the case in a monoplant. In the latter case, the piping can be clearly defined by invariable quantities, such as the corrosiveness of the medium (choice of material), volumes conveyed, or required or allowed flow rate (pipe diameter), as well as pressure of the medium (wall thickness). In a multiproduct plant, each of these requirements is described by a range, which at the time of planning may not be well defined [5.1]. Another question which may arise is whether the piping should be permanently laid, or, if they are only temporarily needed, if hose assemblies should be fitted. For this reason, construction elements such as adapters, blinds, hoses, couplings, and so forth are found more often in multiproduct plants than in monoplants. The design should allow, as far as economically possible, for all eventualities. In other words, the margin of play in such a construction is greater than that of monoplants.

An example of this is the design of the accompanying heating for piping. Fig. 5.1 shows different possibilities. Accompanying heating in the form of an accompanying pipe is only suitable for simple tasks. If the product temperature may be substantially lower than that of the heating medium, and a cold pipe doesn't need to be heated rapidly, this is an inexpensive alternative. Double-mantle accompanying heating systems, shown as 2) and 3) in Fig. 5.1, are more expensive, but they also allow substantial heat transfer to the piping. If, as in 3), the flanges are also included in the accompanying heating, melts may also be handled only just above their melting points. The flange pairs of especially narrow pipes (up to DN 50) cause cold spots where the product can solidify and block the pipe. If, at the time of planning, the melting- or crystallization points of the materials to be transported in the piping at a later stage are unknown, it may benefit the product-assortment flexibility if more is invested in a better accompanying heating system, one that incorporates the flanges.

One should choose high-quality materials if there is a possibility that highly corrosive media will also be handled in the plant. In multiproduct plants, the design of the piping is often determined by the units connected by them. The material used, the diameter, and the pressure-handling ability of the piping are therefore selected according to the maximal demands placed by the connected apparatus and machinery. One is then equipped to handle every procedure that may possibly be run in the connected plant parts. Stainless steel mixing vats with a specified pressure of 6 bar will, for example, be connected with PN 10 (nominal pressure rating 6 bar) stainless steel piping. If the vat is connected to a circulation pump with a maximum delivery rate of 6 m^3 h^{-1}, the dimensions of the branching connection to another apparatus will be such that the total throughput can be handled. Lined piping will be chosen to handle as wide a range of requirements as possible. If it is not

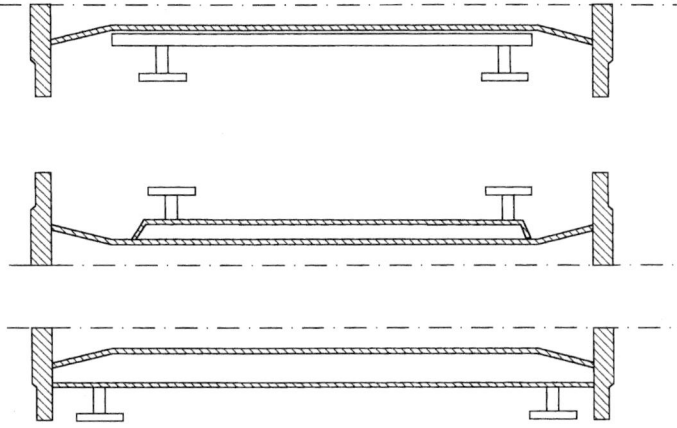

Fig. 5.1 Various possible accompanying heating systems for pipes. 1) Accompanying tube, 2) double mantle, 3) double mantle with incorporated flanges

known whether piping, especially of large diameters, should be capable of operating under vacuum, one should use glass-lined rather than plastic-lined pipes.

Piping supplying energy, such as cooling water, vapor, and cooling brine, as well as feedstocks used all over the plant, are permanently installed in multiproduct plants too (piping lanes). With permanently installed pipelines, one should take care, especially in explosive areas, that the pipeline sections don't get too long. A flange placed approximately every 3 m along a pipeline has proved to be useful should alterations become necessary; in such a case, the flanges can easily be opened or released. In this way, welding and cutting operations within the plant can be avoided.

There are well-proven constructions and designs for cases where the piping often needs to be altered. Adapters are useful where changes in the connections between pipes or plant parts are needed between different product campaigns. Alternative material-flow possibilities can be realized with simple curved adapters. Such possibilities are described in Section 3.4, and shown in Fig. 3.17. Ready-made piping parts (modular piping), which can be obtained directly from the stock room, can substantially reduce the alteration time and subsequent downtime. This type of standardization is common in the case of glass and lined pipes. There are substantial advantages to standardization, especially where construction elements such as valves and flanges are involved (see also Sections 5.3 and 5.4).

It is also recommended that the cleanability requirement is followed when pipelines are designed. The piping in multiproduct plants usually contains more discharge- and cleaning ports than in other plant types. Where possible, the piping should be constructed to run at an angle, to make complete drainage possible. The example shown in Fig. 4.21 for the design of a pump illustrates how careful placing of rinsing ports makes it possible to disconnect and rinse specific sections of the piping during product changes or maintenance work.

5.2.2
Hose Lines

A hose line is a length of hose with hose fittings on both ends making it ready for use (see Fig. 5.2). In principle, hoses can be divided into those made of elastomers and thermoplastics and those made of corrosion-resistant (stainless) steel (metal hoses). Hose lines are used when:

- Connections between pipelines are only needed for limited time periods
- Stationary and mobile apparatus need to be connected to each other
- A connection point needs to be used in different places

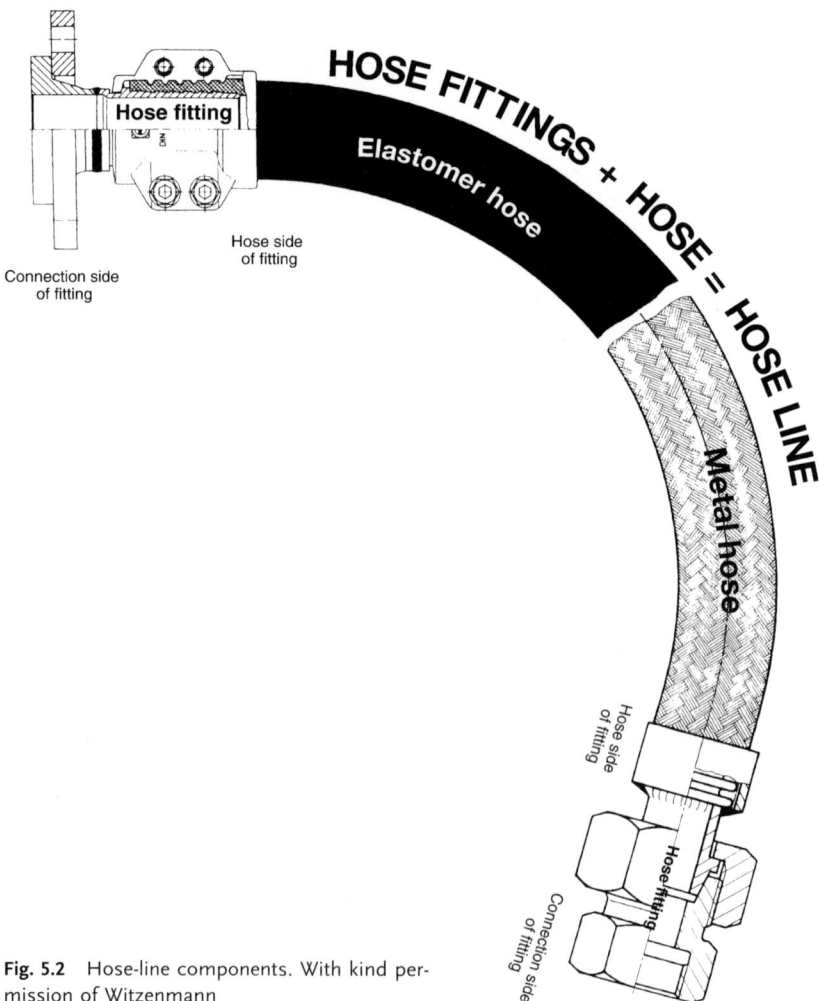

Fig. 5.2 Hose-line components. With kind permission of Witzenmann

Hose lines endow multiproduct plants with considerable structural flexibility (see Section 1.2). With hose lines, the new plant structure needed for a different product campaign can be realized rapidly in modular multiproduct plants; one simply places the movable modules in the required places and connect them with the permanently installed plant parts (e.g., agitated reactor vessel, distillation column with condenser) or with one another by using hoses and standardized adapters (hose fittings). Such assembly is very simple. In multiproduct plants with pipeline manifolds, the permanently installed apparatus can also be connected to each other quickly and flexibly through the hose couplings [or, alternatively, adapter pipes (pipeline elbows)] found on the manifold plate.

Hose lines are also used for loading and discharging containers (drums, flasks, vats, vessels, etc.), even tanks. Hoses are used in such cases because one of the connection sides needs to be movable. The flexibility inherent in hose lines can also, in principle, be used for absorbing vibration, elongation, bending, and lifting movements. For containers that need to be weighed while connected, that is, with the hose lines intact, the hoses can be used for decoupling forces, so that the measurements obtained are true.

The specifications of hose lines and the handling of hose lines are dealt with in numerous instructions, technical rules, regulations, and norms; German examples of these are: DruckbehV (pressure vessel decree), VbF (decree regarding flammable liquids), TRbF (technical regulations for flammable liquids), WHG (water management law), VVAwS (management instructions regarding decree for plants handling possible water pollutants), DIN 2817, DIN 2823, and DIN 2825 to DIN 2828. The Association for Industrial Safety of the German chemical industry has produced a pamphlet dealing with the safe use of hose lines [5.2].

Decisive for the safe use of hose lines is its capacity regarding pressure, temperature, and resistance to the substances carried through it. A multiproduct plant will therefore usually have a wide selection of hose lines available, capable of coping with a variety of substances and operating conditions. There are also elastomeric hoses available as so-called multipurpose hoses, which are capable of covering a wide range of applications. Such hoses contain, for example, an inner layer consisting of cross-linked polyethylene (PE-X) or fluorinated plastics [e.g., PTFE, PFA, FEP (fluorinated ethylene–propylene)]. They are resistant against a wide range of products and solvents. Pressure hoses consist of an inner layer, strength-supporting material, and an outer layer (see Fig. 5.3). Aspirating pressure hoses additionally contain a spiral between the supporting-material layers as cross-section stabilizer. The inner layer, which should be inert and impermeable towards the substances transported through the hose, is seamless. The outer layer should protect against mechanical damage from the outside and short-term external chemical influences. Metal hoses, mainly of stainless steel, are corrugated and are surrounded by a covering. They are preferably used for the transport of chemical substances in gas-, vapor-, or liquid form [5.2]. The applicability of metal hoses can be extended through the use of high-quality materials, such as Monel or Hastelloy C steel. An advantage of metal hoses is that they can be manufactured as double hose lines which can be heated or cooled (Fig. 5.4). Hose lines for trans-

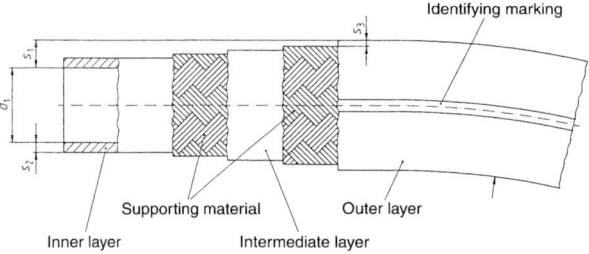

Fig. 5.3 Construction of a pressure hose

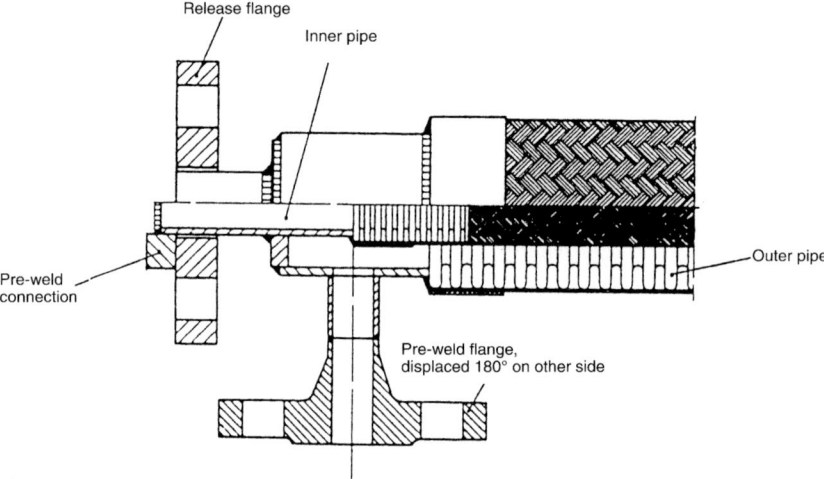

Fig. 5.4 Double metal hose line; can be heated or cooled

porting flammable substances, dust-explosive solids, or electrostatically chargeable materials in explosion-dangerous areas need to be electrostatic conductors. For elastomeric hose lines it is therefore important that the connection between the hose fittings and the hose is also conductive.

One distinguishes between two sides of the hose fittings: the hose side and the connection side (see Fig. 5.2). For securing the hose fittings to the hose, the only attachments that are allowed in the case of elastomer hoses are clamp grips, press grips, and, in exceptional cases, vulcanized hose fittings [5.2]. Securing hoses with hose clamps, clamping tape, assembling wire, and so forth is not allowed. Hoses consisting only of PTFE complete with release flange and lap-joint (flared tube end) is also available from manufacturers. The advantage of this version is that the possibly corrosive transported substances don't come into contact with the metallic surface of the hose fittings. Hose fittings should always be welded onto the hose side of metal hoses; the welding joint should be seamless, that is, without burrs or cracks. A whole range of hose fittings is available for attachment to the connection side of the hose lines (see Section 5.3); these should be chosen accordingly to the operating conditions. It should be ensured that the hose lines cannot mistakenly be exchanged for different application areas and purposes (substances and operating conditions). Different application areas and purposes can include, for example, different chemical substances (feedstocks, products, solvents), steam, heating water, water, brine, nitrogen, vacuum, waste gas, compressed air, control air, breathing air.

If product loss is to be avoided, hose lines, just like pipelines, should run in a way that enables complete drainage (a downward angle, no blind ends). With corrugated hoses, running at the correct angle does not always accomplish this. In critical cases, as in pharmaceutical production or under GMP conditions, hoses with a smooth inner layer (e.g., PTFE) is therefore preferred. Between products, hose lines are cleaned at a separate cleaning station after being dismantled, in contrast to pipelines, which are cleaned while still fitted.

5.3
Flanges, Couplings, and Seals

In multiproduct plants that should be cleaned in a closed state (cleaning in place, see Section 4.3), that is, without plant parts and piping being dismantled, a basic requirement is that the number of connections should be reduced to a minimum, because each joint with its accompanying seal represents potential dead volume; this has a deleterious effect on successful and consistent CIP cleaning and can also lead to cross-contamination upon product changeovers. Metallic pipelines are therefore welded together where possible. Well-suited for this purpose is the so-called orbital welding method, with welding tongs and rotating burner. Very smooth welding seams, no bumps or grooves, result from this automatic welding method, so that even a completely horizontal pipe can be drained completely. Special parts can be used with glass-lined steel, steel/PTFE, and glass tubing, to avoid the use of flanges and seals. An example of such a special part is the distribution tube used for connecting different substance inlets to a socket of an agitated reactor vessel (Fig. 5.5); it is one solid piece and does not consist, as is usually the case, of flanged-together T-pieces.

Where connections need to be taken apart, flanges, screw-, and ring-tension joints and couplings with suitable seals are used; they represent the connection components between the apparatus, machinery, and fittings on the one hand, and pipes and hose lines on the other. They also connect lines to each other, either directly or via pipeline manifolds. They are needed everywhere where separation is a requirement for the flexible reconstruction of a multiproduct plant for product changeovers (see structural flexibility, Section 1.2) or for cleaning, maintenance, servicing, and inspection purposes.

Apart from the basic requirements placed on the connection components of multiproduct plants, namely suitability for the chosen operating conditions (pressure, temperature, flow rate, etc.) and substances used (corrosion-resistance, im-

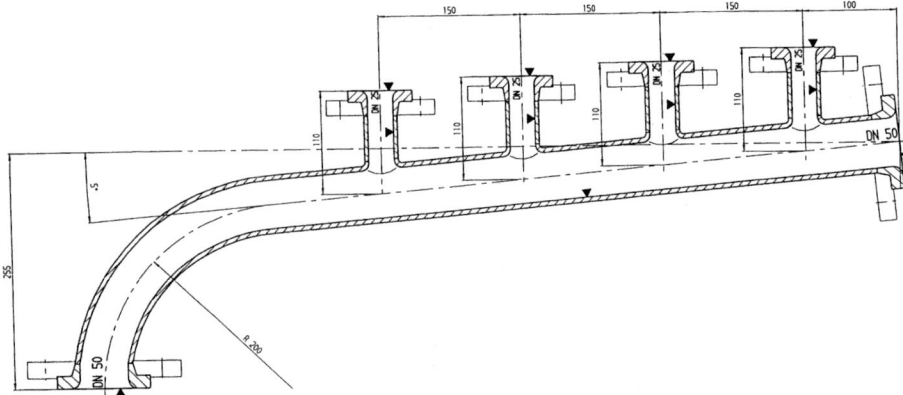

Fig. 5.5 Specially designed distribution tube (glass-lined steel). With kind permission of Düker

penetrability, etc.) as well as being leakproof, there are two more important points to note:

- If a joint needs to be taken apart and put together often, this should be possible in a simple, fast, and secure way (preferably without tools)
- Joints that are rarely taken apart, should be especially easy to clean while fitted

Where CIP cleaning is required, flange connections that have no or only little dead volume are recommended for metallic piping (see Fig. 5.6). An elastic O-ring seal is placed in a ring groove and closes tightly to the product-space side. Because of the form-fitting construction and the pre-stressing of the O-ring seal, neither product nor cleaning agent can run into the seal space. The area of the seal that comes into contact with the product is reduced to a minimum. The defined compression of the O-ring is ensured by an axial metallic contact (stop) be-

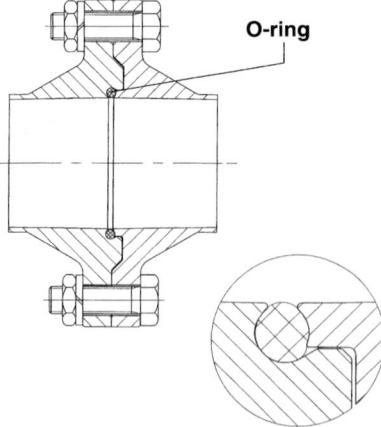

O-ring

Fig. 5.6 Dead-volume-free flange joint. With kind permission of Guth

hind the seal. In this way, all the remaining forces (attraction, heat-tension forces) have no influence on the sealing element and its sealing effect.

The principle described above can also be used for other joint types, such as welded screw joints and locking-ring joints (see Fig. 5.7)

Technological developments in the area of glass-tubing connections have aimed at reducing the ring gaps at the side of the product, to facilitate CIP cleaning. A newly available alternative to the generally used ball-and-socket joint is the fire-polished conic flat flange with composite joint [5.3]. The fire-polished contact surface and elastic PTFE-covered seal which is almost completely closed to the product side make cleaning considerably easier. The elastic core of the seal allows, within limits, the two planes of the joint to be applied unevenly; this causes some correction in alignment. Jointed seals are also used specially for this purpose (Fig. 5.8). They consist of two PTFE jackets with two O-ring beads with a metallic joint set at the center.

In glass-lined steel pipelines, there has so far been no alternative to the generally used release-flange (removable-top-plate flange) joint with even contact surface and flat seal, whose ring gap in the product space makes cleaning difficult.

Couplings are the most suitable connecting elements wherever connections have to be taken apart and put together again on a regular basis. They are therefore very important components of multiproduct plants, supporting structural flexibility in their role as fittings on the connection side of hose lines (see Section 5.2) and the adapters to pipelines. To avoid confusion, suitable steps should be taken; for example, different couplings and hoses that have been labeled accordingly should be used for different purposes.

The types of connections preferentially used for hose lines are described by the German industrial norms DIN 2817, 2826, 2827 and the industrial standards of the German chemical industry. Example of these are the connections to mobile tanks (tank wagons), couplings with swing levers and claw joints, that may only be used for water (temperature < 50 °C) and compressed air; other examples are fast-release lever joints with PTFE lining for feedstocks (chemicals) and fast-release joints that close on both sides, made of various steel and stainless steel types, used at

Fig. 5.7 CIP-complying pipe connections. With kind permission of Tuchenhagen

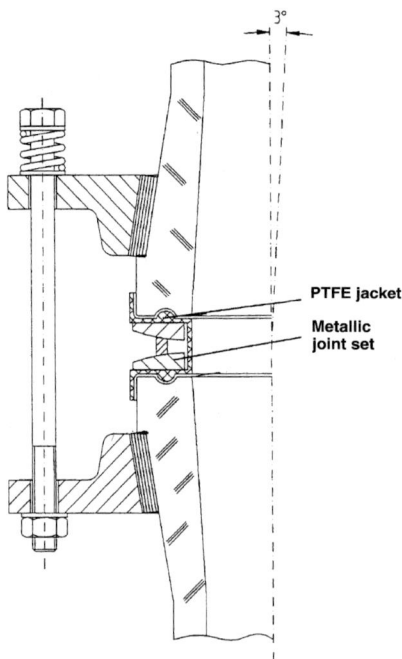

Fig. 5.8 Glass connection. Safety flat flange with jointed seal. With kind permission of QVF Glastechnik

PTFE jacket

Metallic joint set

higher pressures (up to 800 bar). In multiproduct plants with pipeline manifolds, the desired connections are brought about manually or automatically via curved adapter tubes (elbows) and dead-volume-free sliding couplings. To avoid mistaken operation, the connections can be coded or equipped with safety features.

Special coupling designs are required for pipeless plants. Here both connection sides are fixed, that of the mobile container and that of the station where the intended process and technical operation is to take place. Several things need to be connected at the same time, for example, the material transport of the feedstocks, the electrical supply of the agitator motor, heating and cooling by means of a heat-transfer medium, and the signal transmission of the measuring and regulating instruments. For this, an automatic transport system positions the containers within a 2-mm range of accuracy [5.4] at the station. Coupling also takes place automatically. For flammable and toxic chemicals, leakproof joints are used. Lever joints are utilized for steam connections. Joints with inflatable seals are recommended where solid materials are transported. For compressed gas, locking-ring joints are used.

The requirements placed on the seals in multiproduct plants can be summarized as follows:

- Resistance to feedstocks, products, and cleaning agents
- Temperature stability
- Vacuum- and pressure-proof construction (e.g., compartmentalization)
- Dead-volume-free

- Smooth, clean surfaces — wearing out should be minimal
- No color transfer to feedstocks and products should occur
- Aging resistance

To avoid mix-ups in multiproduct operations, a large assortment of seals should be available, and a so-called seal key should be used to determine the standard seals for the required applications (medium, operating conditions).

5.4
Valves

Flexibility is the priority when valves are chosen for use in a multiproduct plant. The casing and the seals of the valves should be made of materials that are suitable for a wide range of applications. Valves made of stainless steel, special materials, and with linings are nowadays available for this purpose. Graphite and teflon-based materials are available for sealing both the spindle and the casing, so that a wide range of possible requirements can be covered. Corrugated-bellow- and magnetically coupled valves are available for special situations. The frequent product changeovers also make cleanability an important criterion in the choice of valves. Valves that comply with CIP and SIP requirements are ideal (see Section 4.3). Valves that have no or little dead volume and can be completely emptied (empty automatically by itself) are well-suited for this purpose. Diaphragm valves are mainly used in this context. If these have been designed to run at a suitable angle, they can be completely discharged. Such a diaphragm valve is shown in Fig. 5.9.

The designer of a multiproduct plant should, because of the frequent rebuilding and alterations taking place in a multiproduct plant, make standardization a priority. If it is not possible to make do with a limited number of valve types, which is preferable, one should then at least stick to using a valve of defined length and flange type for each nominal width. This should be the case for both manually and automatically operated valves. Only valves that have a standardized connection for a drive should be used; this way the drives are then also widely standardized.

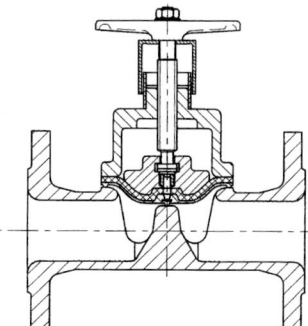

Fig. 5.9 Schematic representation of a diaphragm valve. With kind permission of Düker

If regulating valves are used, an as wide as possible regulation range is useful. A regulation range of 1:20 is by now regarded to be standard, although even a regulation range of 1:50 can be too narrow in multiproduct plants. In such a case, a regulator with exchangeable trimming kits is useful (see more on this in Section 7.1). Where possible, regulators that close tightly should be used (soft-seat models), since, depending on the process, it may be used either for regulating or for switching on/off.

5.5
Pigging Technology

The flexibility required from multiproduct plants also leads, as described previously, to high demands being placed on the cleaning of product-carrying pipes. Every time the product is changed, the unit needs to be completely drained, to free it of all residues, and to completely clean it, so that product loss and cross-contamination of products (see Sections 4.2 and 4.3) are avoided. An effective and economical process for product transport and for emptying and cleaning of pipes is that of pigging technology.

A "pig" is a special contoured plug that fits into the inside of a pipe and is used for pushing the pipe contents out; the pig itself is moved by a propellant (gas or liquid, e.g., air, nitrogen, water, cleaning agent, or even product) through the pipe. With this technique, several different products can be transported separately from each other through the same line. An almost complete evacuation of the pipe becomes possible.

Pigging technology originated in the mineral-oil industry [5.5], where pipelines have been cleaned with metal-fit bodies for decades. Since then, pigging systems have also been used in the chemical, paint, and, increasingly, the food, cosmetic, and pharmaceutical industries. Pigging is often used as an economically viable process for transferring various products through the same pipelines [5.6]. The versatility of the branching and connection possibilities of piggable pipelines facilitates the construction of very flexible filling and transporting plants.

The advantage of pigging systems, apart from the reduction of product losses and quality assurance through the avoidance of cross-contamination, is the reduction in cleaning-agent use (especially with long pipelines) and waste-water contamination or waste-combustion costs. If the conventional, time-consuming draining and cleaning procedures are no longer needed, expensive production downtimes are reduced. If the number of pipes and valves that need to be installed are less as a result of pigging technology, the required space is also less, with lower investment costs resulting. A further advantage is the possible full automation of pigging systems, and therefore fewer possible operating errors. Balancing the advantages of the pigging system is the increased maintenance required by the pigging fittings and electrotechnical facilities [5.6]. This all depends on the specific properties of the product being manufactured. The pig itself also needs to be exchanged from time to time.

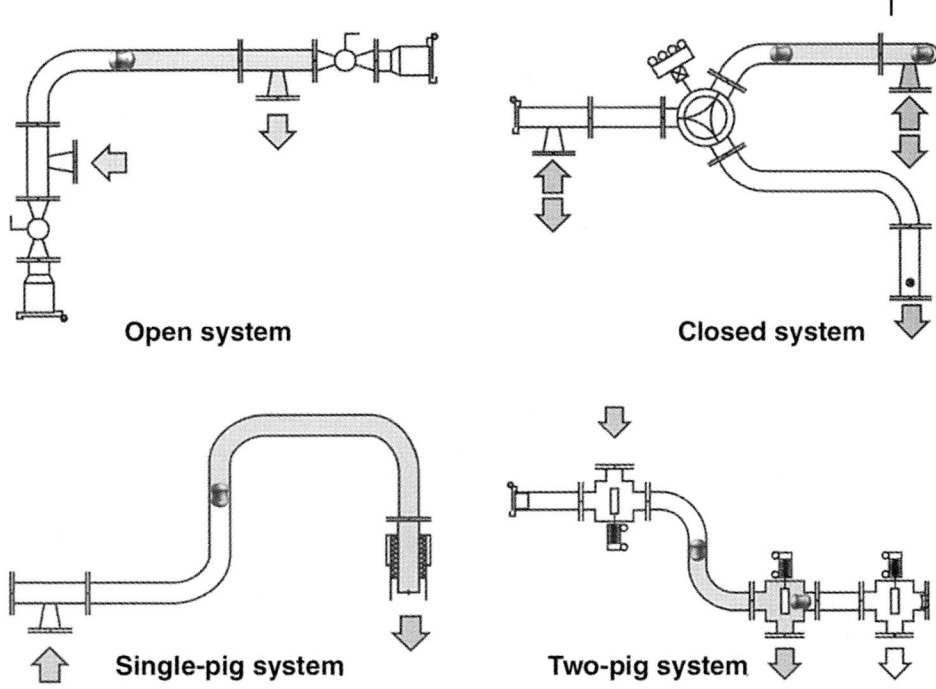

Open system

Closed system

Single-pig system

Two-pig system

Fig. 5.10 Pigging systems. With kind permission of I. S. T. Molchtechnik

The choice of a suitable pigging system depends on the number and properties of the various products that are to be transported, the cleaning requirements, and the investing cost limits. One distinguishes between open and closed and single- and two-pig systems (Fig. 5.10).

In an open system, the pigs are forced into and expelled from the system as required. The system works on a one-way principle, that is, the pigs are conveyed in one direction only, from the launching- to the receiving station. At the end of several pumping operations, the last pig is pushed with compressed air or nitrogen to the receiving station.

In a closed system, the pigs remain in the system for their entire operational life; they do not need to be constantly forced into and expelled from the pipe. After they have completed going through the whole line, they are registered at the receiving station, and are then pushed with compressed air or nitrogen back into the launching station, where they remain until the next pumping operation. The closed system is suitable for a large number of applications, for example, branched systems with three-way pig valves.

In a single-pig system, only one pig is in the line at any time. It goes into action when, after a pump cycle, product residues need to be removed from the pipe.

The two-pig system is one of the most commonly used ones. Depending on the application, either product or cleaning agent is fed into the system, between the two pigs. In the first case, both pigs are found at the beginning in the launching station. When the pumping process begins, the first pig is pushed into the line and the product is then pumped in after it; this results in the air being pressed out of the line. This process can, for example, be used for a filling process that needs to proceed without foaming. The product flows past both pigs and out of the line. Once the conveying action is completed, the residual product can either be pushed back into the tank by the first pig, or be pressed out of the line into a tank wagon, for example, by the second pig. For this, the pigs are propelled by cleaning agent. With compressed air, the pigs may subsequently be returned to the launching station. In the second case, the so-called tandem method, the first pig, propelled by cleaning agent, pushes the product out of the line. For this procedure, a defined amount of cleaning agent is contained between the first and second pig. The pigs can be pushed into the receiving station with the aid of rinse water, and back to the launching station with the aid of compressed air.

Great demands are placed on the pig itself (see Fig. 5.11). Pushed to make as much contact with the tubing walls as possible, it needs to effect the highest degree of cleaning while wearing out as little as possible. Openings in the pipes should be passed and curves in the tubes should be followed with minimal friction. The material they consist of should be resistant to the product and cleaning agents. Depending on the area of application, FDA-approved pigs (NBR-L, EPDM-L, silicon, natural rubber) or chemically resistant pigs (e.g., Vulkozell, NBR, EPDM, FKM) can be used [5.5]. The almost universally resistant PTFE is unfortunately not suitable, because of its low elasticity. The applicability of pigging technology in multiproduct plants is therefore limited by the resistance of the material of which the pig is constructed.

A permanent magnet is found inside the pig; this is used for detection in the launching and receiving stations of automated systems or in the pipelines and pigging fittings.

If cleaning is to be efficient and the pigs are not to be damaged, piggable pipelines should be smooth, of uniform diameter, and without barriers; they should also be well-supported, because of the resulting rocking motion. Flanges should

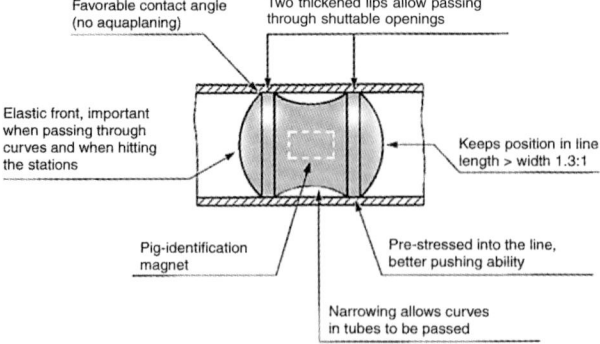

Fig. 5.11 DUO pig. With kind permission of W. Mühlthaler [5.6]

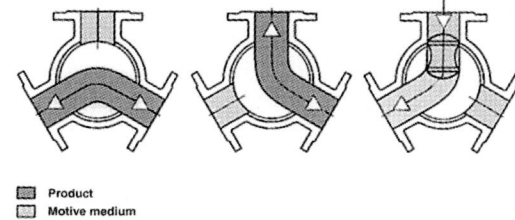

Fig. 5.12 Three-way valve, dead-volume-free. With kind permission of W. Mühlthaler [5.6]

☐ Product
☐ Motive medium

be exactly centered. The seals should have the same diameter as the pipe and should not intrude into the inside of the tubing. Flanges with bordering O-ring seals (see Section 5.3) are available for this purpose. Welding seams should be without burrs (orbital welding method) or should be polished from the inside. Valves, measuring instruments, and fittings should not intrude into the available pipe cross-section. Instead of standard T-pieces, piggable T-pieces with special T-branches with reduced inside diameters should be used, so that pigs can move through them into the straight passage. If the surface roughness of the tubes is low (e.g., $R_a = 0.8\,\mu m$), cleaning is more efficient and the pig undergoes less wear and tear. The cleaning efficiency can be improved by attaching the piggable pipelines to a CIP cleaning facility by means of special valves.

Apart from the already mentioned launching and receiving stations, there are several other special pigging fittings, for example, T-ring valves with three functions: piggable T-piece, product isolation valve, and pig stop; loading lances for application in piggable filling plants; pig loading and unloading stations; and manifolds. Pipeline manifolds add to the structural flexibility of multiproduct plants and reduce the numerous lines required between tanks, reactors, and filling stations to a minimum. There are several manifold systems available for use with piggable pipelines [5.7], for example:

- Three-way valve, dead-volume-free (see Fig. 5.12)
- Rotary manifold (connects one or two supplies with up to 18 outlets)
- Modular multidirection manifold (connects 12 inlets to a maximum of 4 outlets)
- Full system manifold (maximum 20 inlets can be connected simply to 20 outlets; see Fig. 5.13)

With the modular principle, each pigging system can be tailored to the application at hand. Pigging technology is limited only by the technical possibilities and the economic efficiency of the investment [5.7].

Fig. 5.13 Full system manifold. With kind permission of I. S. T. Molchtechnik

5.6
Interlocking Systems

A characteristic of multiproduct plants is that the available piping can be used for very different substances and for very different routes. This property, desirable for its benefits to flexibility, is also associated with the risk of undesired substance flow, substance mixing, and substance discharge. Various organizational steps can be taken to avoid such problems. Precautions going as far as having the piping connections and valve settings checked by two independent persons are, for example, practiced. Such precautions may be supported by technical and automated problem solving; this requires less operating input, yet offers substantial security. Well-proven in this context are various interlocking systems. The simplest systems are purely mechanical, and separate only two or three different material or energy streams from each other. Such bolts may be equipped with valves or ball valves. Fig. 5.14 shows a manually operated bolt for energy; this bolt is supplied with ball valves. The hand lever is displaced from the normal setting by 90° and is cropped. With such a bolt, the inlet or outlet of heating vapor, cooling water, or cooling brine can proceed according to need. With this system, it its impossible to discharge the cooling brine into the cooling water or the condensate. Not discharging the cooling water and condensate into the sewerage system contributes to avoiding pollution of effluent with glycol-containing brine. The same bolting mechanism is also suitable for preventing the undesired mixing of two or three product streams.

The flexibility required from multiproduct plants is facilitated even more by locking systems. This purely mechanical system also operates with valves or ball valves. The commercially available valves are equipped with locks; oppositely oper-

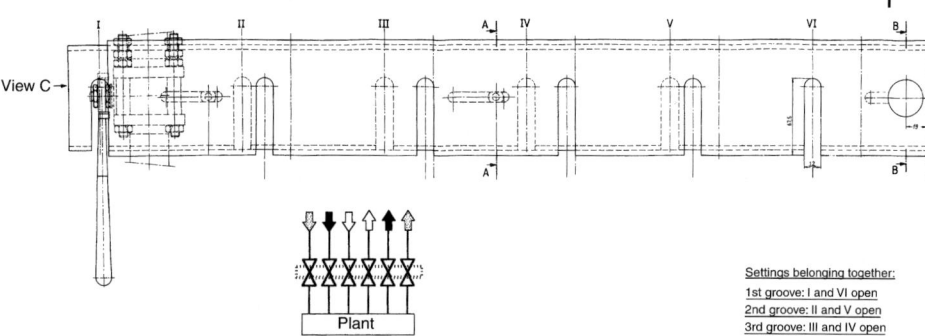

Fig. 5.14 Interlocking system for brine, cooling water, and vapor

ating valves are operated with the same key. In this way, valve 1 can, for example, only be opened if the key to valve 2 has first been removed, which is only possible if valve 2 is closed. This system is also suitable for further upgrading to valves that are far apart. TÜV-tested and -approved (TÜV, the German Technical Inspection Association) versions of this system are available.

Apart from specific substances being safely locked off, the way the unit is switched at any time is also of interest. Solutions to this are integrated valve and indicator systems, which make the information about the latest switched position of a plant available to the central measuring station. Even the information on how hose lines in switching yards are switched are available by this method [5.8] (see also Section 3.4 and Fig. 3.19).

An elegant method of interlocking, especially in technical situations where safety does not play such a major role, is with the aid of process control technology. For this, a distributed control system should be available in the multiproduct plant and the locking of the valves should be operated by remote control. Here specific marginal conditions that determine whether certain switching positions are allowed or not are part of the system. Where distributed control systems are used in conjunction with a recipe control package, the locking strategy for the system becomes part of the recipe.

5.7
References

[5.1] THIER, B. *Chem.-Ing.-Tech.* **1973**, *45*, 480–485.

[5.2] BG Chemie *Schlauchleitungen, Sicherer Einsatz* **1995**, 9, pamphlet T002.

[5.3] *Chem.-Anlagen* Verfahren **1995**, *28(7)*, 104–106.

[5.4] TADAO, N. *Chem. Eng.* **1993**, *(June)*, 102–108.

[5.5] LAGONI-OPITZ, C. *ZFL* **1996**, *47(3)*, 16–19.

[5.6] MÜHLTHALER, W. *Chem.-Ing.-Tech.* **1995**, *2*, 171–175.

[5.7] KLUDAS, H.-D. *Chemie-Technik* **1995**, *5*, 54–56, offprint.

[5.8] Anonymous *Process* **1995**, *7/8*, 56.

6
Materials

6.1
Introduction

In contrast to single-product plants, where the physical and chemical operating conditions of the individual plant units are very similar, these parameters are less focused in multiproduct plants. This invariably has an influence on the choice of the materials used for the plant components, which should be able to withstand very different conditions, need to fulfill very specific product requirements in some cases, and simultaneously need to be easy to work and economical. In Fig. 6.1, the numerous requirements and considerations that need to be accommodated or at least clarified just choosing materials for a conventional chemical plant are set out. Choosing suitable materials for a multiproduct plant, with its great variety of different requirements, obviously presents an even more complex problem. To address this problem properly, one should ideally already in the planning stage be aware of the range of physical conditions, such as pressure, temperature, rates of pressure changes, flow rates, presence of multiphase streams, and the associated erosion or, possibly, the cavitation properties. Further, just as important is a knowledge of the range in the corrosivity of the media, including pH values, quantities and types of anions and cations in the solution, as well as the electrochemical corrosion potential. Since it is generally not possible to know all this beforehand, one should choose "broad-spectrum-resistant" materials for multiproduct plants, that is, materials that are sufficiently stable should the parameters change. Special product requirements, such as strict product-purity requirements, should also be accommodated.

This large number of determining factors for the choice of materials in multiproduct plants means that recommendations that are generally applicable or are directly pertinent to the individual reader are not possible. Therefore, in this chapter, the application areas and properties of various materials will be focused on instead. If the application requirements correspond to the described application areas of a specific material, this material would be a possible candidate for that purpose. From this range of possible candidates, the most suitable material can then be chosen on the basis of corrosion experiments and by consideration of further requirements and factors (see Fig. 6.1).

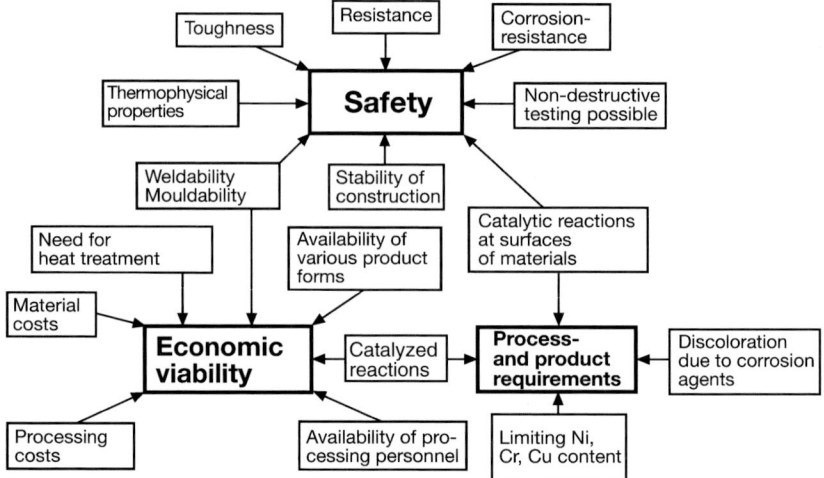

Fig. 6.1 Requirements and considerations when choosing materials for chemical applications

6.2
Strength Properties and Application Temperatures of Materials

The construction of the plant components is based on the strength properties of the materials. For this, a prerequisite is that the materials have sufficient toughness and workability, otherwise, in complex constructions, unavoidable notches can lead to brittleness at low nominal stresses. Because the strength and toughness of materials are generally inversely proportional to each other, and the strength is higher at lower temperatures, it follows that the toughness of most materials decreases at low temperatures. This can, in certain cases, lead to a lowest application temperature limit. In the German pressure-vessel code (AD/W 10), the lowest application temperature of the metallic materials approved for use in pressure-vessel construction is linked to the mechanical stress placed on the part.

For the special metals and plastics not covered here, the lowest approved application temperatures are given in Tab. 6.1.

In Fig. 6.2, various iron-based alloys that can be used for the handling of liquid gases are shown.

The upper temperature limits to the application of materials, in contrast to the lower limits that are generally set by the loss of toughness, are determined by the

Tab. 6.1 Lowest application temperatures for special metals, some plastics, and enamel

Special metals		Plastics		Enamel
Ti, Zr, Ta	*PP, PVC*	*PE, PVDF*	*PTFE*	
–10 °C	0 °C	–40 °C	–60 °C	–60 °C

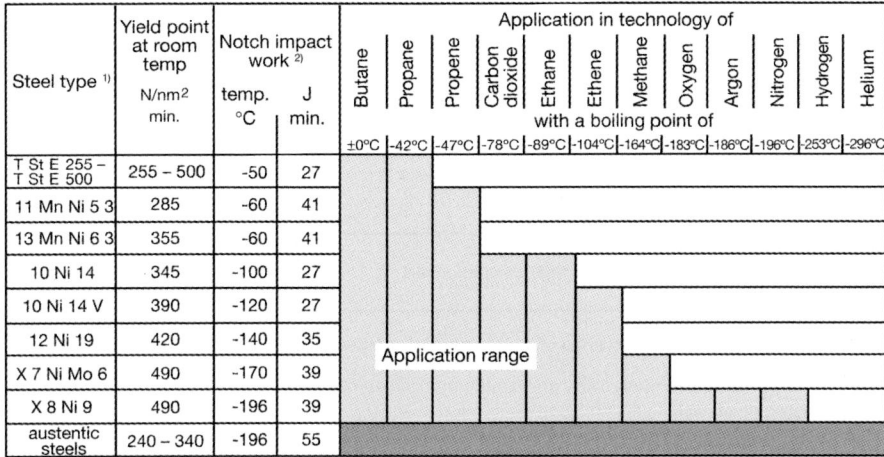

Steel type [1]	Yield point at room temp N/nm2 min.	Notch impact work [2] temp. °C	Notch impact work [2] J min.	Application in technology of											
				Butane ±0°C	Propane -42°C	Propene -47°C	Carbon dioxide -78°C	Ethane -89°C	Ethene -104°C	Methane -164°C	Oxygen -183°C	Argon -186°C	Nitrogen -196°C	Hydrogen -253°C	Helium -296°C
T St E 255 – T St E 500	255 – 500	-50	27												
11 Mn Ni 5 3	285	-60	41												
13 Mn Ni 6 3	355	-60	41												
10 Ni 14	345	-100	27												
10 Ni 14 V	390	-120	27												
12 Ni 19	420	-140	35												
X 7 Ni Mo 6	490	-170	39												
X 8 Ni 9	490	-196	39												
austentic steels	240 – 340	-196	55												

1) Chemical composition   2) ISO impact-notch sampler; average of three individual tests.

Fig. 6.2 Application range of steels with low-temperature toughness in liquid–gas applications [6.1]

thermal softening of the materials, solid-phase reactions, such as precipitation processes, and surface reactions, such as excessive oxidation processes. Where unalloyed steels are used, even in components not under mechanical strain, such as heat shields, high-temperature use in an oxidizing dry atmosphere is confined to a maximum temperature of 570 °C. Above this temperature, apart from the iron oxide modifications hematite (Fe_2O_3) and magnetite (Fe_3O_4) forming, of which especially magnetite leads to the formation of a protective covering layer, wustite (FeO) also forms. Because of the higher density of lattice imperfections in a wustite layer, this oxide develops relatively fast, thereby causing considerable metal loss.

Materials with high chromium content, on the other hand, form chromium oxide layers or chromium–iron–oxygen spinels on the metal boundary; the protective effect of some of these range up to 1000 °C. Even higher temperatures are possible with special steels, additionally alloyed with silicon or aluminum, which form silicon- and aluminum-rich oxides as protective layers (**SiCrOMAl** steels). Nickel- and cobalt-based alloys are alloyed similarly (e. g., with Cr, Si, Al), to obtain high-temperature applicability. These materials usually have the advantage over iron-based alloys that they are more resistant.

The special alloying elements, mostly small amounts of cerium and yttrium, used most often in modern high-temperature alloys mainly function by causing better adhesion of the protective oxide layers. Water vapor in the atmosphere can cause a substantial reduction in the upper temperature limits.

An overview of the application temperature limits of various high-temperature materials is given in Fig. 6.3.

Fig. 6.4 shows the maximum application temperatures of some materials that are of interest where chemical-corrosion resistance is necessary.

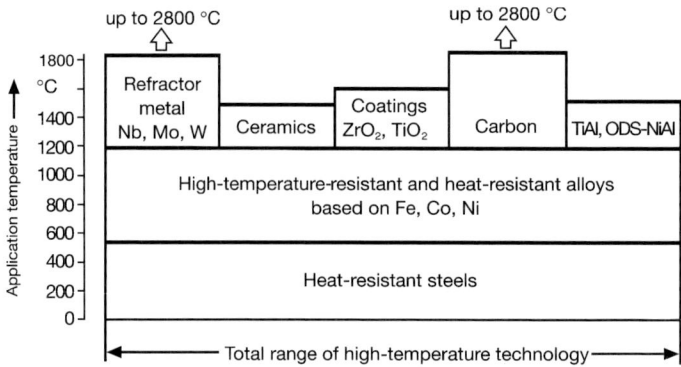

Fig. 6.3 Application temperatures and application areas of high-temperature-resistant materials

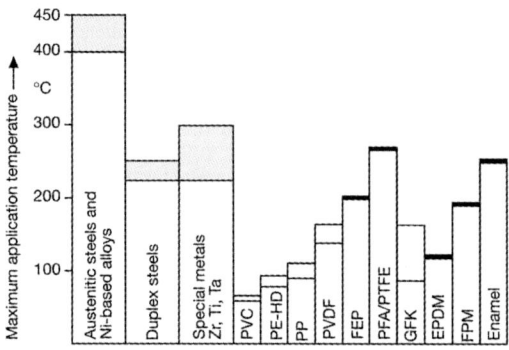

Fig. 6.4 Upper temperature ranges at which corrosion-resistant materials can be used continuously

The temperature-limit values for the metallic materials are taken from the VdTÜV (Technical Inspection Association of Germany) materials leaflets and the AD (Study Group on Pressure Vessels) leaflets. With the corrosion-resistant steels and the nickel-based alloys, there is a danger after continuous use at higher application temperatures that they will build embrittling precipitations, which also lead to reduced corrosion resistance. With some of the special metals, a strongly temperature-dependent softening occurs above the borderline temperatures. There is a general danger with all special metals that at above 200 °C, gas absorption (O, N, H) of the metal increases. The use of enamel at high application temperatures is limited by the use of PTFE seals. Please note that for all the upper operating temperature limits given in Fig. 6.4, the effect of the media has not been considered. This means that, in some cases, when the materials are used in conjunction with aggressively chemically corrosive media, operation may only be permissible at considerably lower maximum application temperatures.

The strength properties of materials are not only important with regard to the dimensioning of the plant units, they also determine the behavior under conditions of erosion and wear. In solid suspensions, for example, although high strength is usually advantageous, the situation becomes very complex when corrosive attack is also a factor, and should be very carefully evaluated. Also, in many cases there may be no room for play with regard to the dimensions of the construction; this may be for reasons of weight (structural considerations), based on processing considerations, or on process engineering grounds (e. g., heat transfer reasons). In such cases, materials with sufficiently great strength and, where required, material combinations such as lining or cladding should be used; there should be a functional separation between the lining or cladding material guaranteeing the corrosion resistance and the strength-determining base material.

In Fig. 6.5, ranges for the tensile- and yield-strength limits at 25 °C for various groups of materials are given.

In contrast to iron and nickel alloys, whose tensile- and yield-strength limits decrease by approximately 25% to 30% as the temperature increases from 25 °C to 200 °C, this reduction may be as great as 40–50% for some special materials. It should additionally be considered that already above 150 °C for titanium and above 170 °C for zirconium, it is no longer the yield strength, but the creep strength $Rm_{100000h}$ that is relevant for the design. For plastics, this is usually already valid at design temperatures above 20 °C. This means that the special metals titanium and zirconium, as well as plastics, in contrast to the iron- and nickel-based alloys, have limited tolerance toward changes in the application temperature. When such special materials are combined, for example, as cladding, with steel, one should then take into account not only the not inconsiderable differences in coefficients of thermal expansion, but also the additional limits set by these.

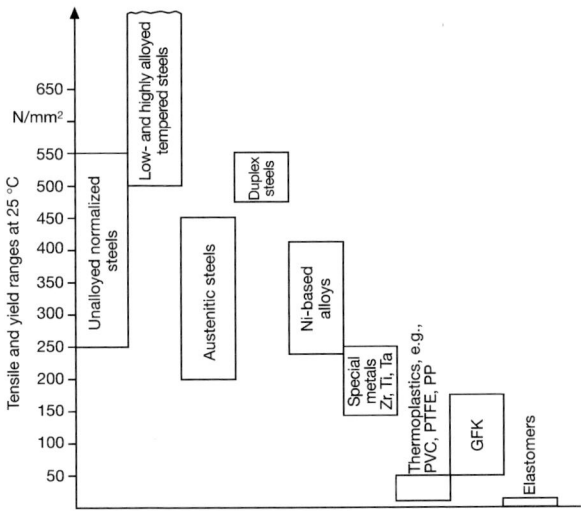

Fig. 6.5 Tensile- and yield-strength limits of various materials at 25 °C

6.3

Types of Corrosion and Criteria for Resistance

In the following section, the corrosion properties of materials in aqueous solutions acting as electrolytes will be dealt with. Because extensive corrosion resistance is a fundamental requirement in multiproduct plants, the highly alloyed austenitic steels, the nickel-based alloys, and the special metals titanium, zirconium, and tantalum will be the only metallic materials covered here, in addition to plastics and enamel. The corrosion properties of materials in hot gases are not included.

Conductive, aqueous media can cause a variety of corrosion types in metallic materials. These can be subdivided into general surface corrosion and localized corrosion forms, such as pit formation, the selective corrosion of construction components, as well as stress- and fatigue-corrosion cracking. A metallic material is generally judged to be technically resistant if its surface corrosion rate is below 0.1 mm per year, but the types of localized corrosion mentioned above are not permissible, as the rate at which they develop is usually above this limit. Technically permissible surface-corrosion limits that have been proven to work well in practice are 0.05 mm per year for enamel and even as low as 0.01 mm per year for special materials such as titanium, zirconium, and tantalum.

The limit set to the allowable surface-corrosion rate of the special metals is related to their tendency to form hydrides; the hydrogen required for this is generated by the cathodic reaction of the corrosion process (see Fig. 6.6). As a result of the atomic hydrogen generated, dependent on the material-determined corrosion potential, hydrogen-induced cracking can also pose a threat to high-strength unalloyed steels and welded joints of superior hardness. This is the reason why steels usually employed in the chemical industry have yield strengths below $400\,\mathrm{N\,mm^{-2}}$ at room temperature.

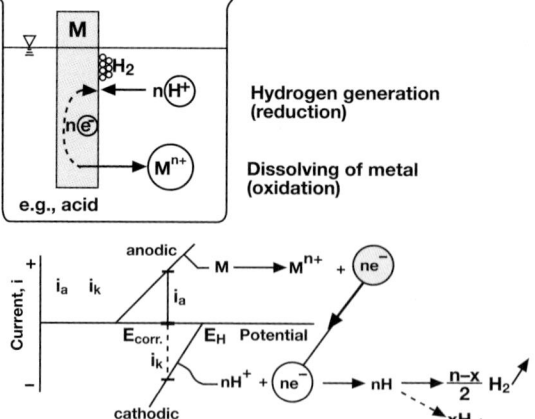

Fig. 6.6 Electrochemical corrosion reactions (at $\varepsilon_{corr} < \varepsilon_{H}$)

For plastics, the resistance criteria are still less clearly defined. Surface corrosion does not often occur with these materials. Here, changes in properties play a more important role; such changes are often caused by diffusion of the media into the material, leading to softening, swelling, brittleness, cracking, and so forth. Evaluating the resistance of plastics therefore requires several criteria to be considered, such as change in mass, dimensions, resistance, and hardness, as well as discoloration, and others. The following values can be used as a general guide: $< 5\%$ change in mass, $\leq 10\%$ change in hardness and resistance, and $\leq 20\%$ change in toughness.

6.3.1
Corrosion Properties of Highly Alloyed, Corrosion-Resistant Steels

Highly alloyed, corrosion-resistant steels are usually iron-based alloys with 22–35% chromium content and 3–7% molybdenum content, whereby, for metallurgical reasons, the latter two elements cannot simultaneously be present at their highest content. These steels are, almost without exception, austenitic steels, with, for stabilizing the austenitic structure, a nickel content of 22–30% and usually also additionally alloyed with nitrogen. These materials are exceptionally corrosion-resistant in acids. Especially in sulfuric acid solutions and in organic acids, but also in strongly oxidizing media such as nitric acid, these materials are superior to the corrosion-resistant nickel-based alloys. The highly molybdenum-alloyed materials are characterized by exceptional pitting-corrosion resistance, also in strongly halide-containing, oxidizing, acidic media. With regard to halide-induced stress-corrosion cracking, however, the highly alloyed, corrosion-resistant steels do not match the resistance of the nickel-based alloys, such as alloy C-4 (material no 2.4610).

In Tab. 6.2, some of the highly alloyed, corrosion-resistant steels and their chemical compositions are listed. Especially the highly molybdenum-alloyed materials are characterized by a great precipitation tendency at high temperatures, and some of them need to be welded not with the same type of steel, but with nickel-based welding consumables. For these steels, the specifications and working guidelines should be followed strictly, to ensure that the properties of plate or pipe materials are also present in hot-formed parts and welded joints. Intermetallic precipitations have a negative influence on both corrosion resistance and toughness.

Especially worth closer attention are the so-called duplex steels, whose name refers to their hybrid structure of ferrite and austenite (ca. 50:50). These steels are characterized by a chromium content of 22–25%, a molybdenum content of 3–4%, and a relatively low nickel content of 5–7%. Duplex steel is highly resistant towards chloride- as well as NaOH induced stress-corrosion cracking. Duplex steel is especially suitable for use under operating conditions in media with pH > 4, and media in which nickel complexation of the more highly nickel-alloyed materials would have been a problem, for example, in hot, ammoniacal solutions. For continuous use at temperatures above 220–250 °C, however, duplex steel is neither suitable nor approved.

Tab. 6.2 Composition of highly alloyed, corrosion-resistant iron-based alloys

Material number or designation	Main alloy content in mass-%				
	Ni	Cr	Mo	Cu	N
Austenitic steels					
1.4465	25	25	2.5	–	0.13
1.4539	25	20	5	1.5	0.08
1.4529	25	20	6.5	1	0.2
1.4563	31	27	3.5	1.2	0.05
1.4562	31	27	3.5	1.2	0.05
654 SMO	22	24	7	0.5	0.5
Alloy 33	31	33	1.6	0.6	0.4
Duplex steels					
1.4462	5	22	3		0.15
1.4410	7	25	3.5		0.28

6.4
Corrosion Properties of Nickel-Based Alloys

Nickel in its pure metallic form (alloy 200, material no 2.4066, alloy 201, material no 2.4068) already possesses excellent resistance in caustic alkali-, halide-, and several organic media. Similarly to iron, it can be alloyed strongly with chromium and even more substantially with molybdenum. Because of this, several nickel-based alloys with chromium and molybdenum with an extremely broad range of corrosion- and temperature resistance could be developed. In Tab. 6.3, the most important alloys for chemical-technical applications are summarized. The wrought alloys listed in this table are characterized by good workability, and especially good weldability with welding consumables of the same type.

Chromium- and molybdenum-alloyed nickel-based alloys are very stable in acids and bases. Especially the highly molybdenum-alloyed materials are additionally resistant against pit formation in acidic, highly halide-containing solutions and are practically immune to halide-induced stress-cracking corrosion.

Especially corrosion-resistant, and therefore often used as a trouble-shooting material, is the alloy C-4 (NiMo16Cr16Ti, material no 2.4610). This and alloy C-276 (material no 2.4819) are the basis of a material family, which consists of the alloys Hastelloy C-22, Alloy 59, Inconel 686, and Hastelloy C-2000. These very closely related materials and material developments, from an alloy-technology point of view, have strongly overlapping corrosion-resistance properties, but have some borderline differences that may be useful in individual cases, although this should always be confirmed by corrosion tests. Note that the more strongly chromium-alloyed materials, for example, C-22, are superior to the less strongly chromium-alloyed materials, for example, C-4, in strongly oxidizing media, whereas the situation is reversed in less oxidizing media. In highly oxidizing media (nitric

Tab. 6.3 Composition of chromium- and chromium–molybdenum-alloyed nickel-based alloys that can be used for chemical applications

Designation	No.	Abbreviation	Main alloy content in mass-%				
			Ni	Cr	Mo	Fe	Cu
Alloy 600	2.4816	NiCr15Fe	Rest	15	–	8	<0.5
Alloy 690	2.4642	NiCr29Fe	Rest	28	–	9	–
Alloy 20	2.4660	NiCr20CuMo	37	20	2.5	Rest	3.5
Alloy 825	2.4858	NiCr21Mo	42	21	3	Rest	2.3
Alloy G3	2.4619	NiCr22Mo7Cu	Rest	22	7	19	2
Alloy G30	2.4603	NiCr30FeMo	Rest	30	5	15	1.8
Alloy 625	2.4856	NiCr22Mo9Nb	Rest	22	9	<3	–
Alloy C-4	2.4610	NiMo16Cr16Ti	Rest	16	16	<3	–
Alloy C-276	2.4819	NiMo16Cr15W	Rest	16	15	5.5	–
Alloy C-22	2.4602	NiCr21Mo14W	Rest	22	14	3	–
Alloy59	2.4605	NiCr23Mo16Al	Rest	23	16	<1.5	–
Inconel 686	2.4606	NiCr21Mo16W	Rest	21	16	<5	–
Hastelloy C-2000	–	–	Rest	23	16	<3	1.7

acid, highly concentrated sulfuric acid), however, the highly alloyed iron-based alloys are preferable to the nickel-based alloys.

In organic acids (also impure), in sulfuric acid (medium concentration range), and in hydrochloric acid solutions, materials like alloy C-types are always suitable candidates. In hydrochloric acid, these alloys are usable over practically the entire concentration range, especially under slightly oxidizing conditions, as long as some surface corrosion at 30 °C can be tolerated. For use in conjunction with acidic substances, the alloys Inconel 686 and Alloy 59 have been shown to be suitable candidates by their good performance in the extremely aggressive flue gas desulfurization washing solutions.

A further important group is that of the pure nickel–molybdenum alloys. The materials belonging to this group are alloy B-2 (NiMo28, material no 2.4617) as well as the further developments Hastelloy B-3 and alloy B-4. These materials are highly resistant over wide temperature ranges in the, rather critical, medium concentration range of sulfuric acid (10–80%), as well as over the entire concentration range of hydrochloric acid, and in some organic acids. Very important if this resistance is to be maintained is the exclusion of oxidizing constituents, such as atmospheric oxygen, Cu^{2+}, or Fe^{3+}, even at low concentrations in the media. Nickel–molybdenum alloys, unlike the chromium-containing alloys, don't form a passive layer. Metal dissolving is therefore the only response to oxidation agents.

The use of alloy B-2 has been associated with problems, especially when used in conjunction with high-purity and therefore iron- and chromium-free melts as well as with thicker walls. After these materials have been hot-formed, welded, and heat-treated, intermetallic precipitations, not detectable by light microscopy, cause embrittlement and corrosion cracking. Judicious alloying with iron and chromium can slow down these precipitation processes.

This strategy has been followed in developing the alloys Hastelloy B-3 and Alloy B-4, without any disadvantage to the resistance resulting. With alloy B-2, deliberate alloying to give an iron content of over 1.5% has also been used, to slow down the unwanted precipitation processes.

6.5
Corrosion Properties of Special Materials

The designation *special materials* is used for the metals titanium, zirconium, and tantalum, as well as for their alloys.

Tab. 6.4 shows the composition of a selection of the most commonly used alloys for technical chemical applications. These materials all have excellent resistance in several aggressive chemically corrosive media. This resistance is supplied by a dense oxide layer that forms on the metal surface; the oxide layer is electrochemically noble, whereas the metals are thermodynamically very unnoble. Titanium stands out with its high corrosion resistance in oxidizing acids, such as nitric acid, but not in fuming, highly concentrated nitric acid. Further uses of titanium are the handling of organic acids, salt solutions, salt brines, cooling-, and sea water. If titanium is alloyed with small quantities of the noble element palladium (0.15–0.2%), the resistance is substantially improved, as the anodic polarization brought about by palladium enhances the protecting oxide layer. The resistance of titanium can be improved to the same extent by addition of strongly oxidizing agents, such as Fe^{3+}, Cu^{2+}, or Cr^{6+}, to the medium. One should be careful with the use of titanium for the handling of alcoholic hydrochloric acid solutions, as stress-cracking corrosion can result. Apart from this, titanium is also not very resistant in strongly alkaline solutions.

Tab. 6.4 Special metals most commonly used in chemical technology

Material/material no.	Details on material	Properties
Ti2/3.7035	≥99.5% Ti	Very high resistance in highly oxidizing media
Ti2Pd/3.7235	Ti2+0.15% to 0.2% Pd	Very high resistance in slightly reducing to strongly oxidizing media
Zr 702	Zr+Hf ≥99.2%, HF ≤4.5%	Soft; very high resistance in acids (except HF)
Ta ES	Pure Ta, electron-beam melted	Very soft; very high resistance
Ta Gs	Pure Ta, vacuum-sintered	Soft, very high resistance
Ta 2.5 W	Ta+2.5% W	Increased strength, very high corrosion resistance

Under extreme, strongly oxidizing conditions, there is a danger of very vigorous reactions with titanium (titanium fires). With red fuming nitric acid and dry chlorine, a titanium fire is already possible at low temperatures. A water vapor content >0.1% in chlorine gas is already sufficient for preventing the danger of pyrophoric reactions.

Zirconium is also not very resistant in strongly alkaline solutions. Zirconium is, however, exceptional in that its resistance is extremely high in numerous inorganic acids (e.g., sulfuric acid up to ca. 65%, at low concentrations also above the boiling point at atmospheric pressure; in high concentrations of hydrochloric acid up to the boiling point; also in nitric acid, not too high concentrations, up to 200 °C) and in boiling organic acids. The presence of strongly oxidizing agents, such as Fe^{3+} and Cu^{2+}, can, in some cases, have a critical influence on the corrosion properties. In hydrochloric acid, small quantities of these oxidation agents (e.g., 10% HCl at 30 °C with 100 ppm $FeCl_3$) can induce pit-formation- and stress-corrosion cracking in zirconium. Similar reports on stress-corrosion cracking in zirconium in hot, technical-grade, concentrated nitric acid have also appeared. Zirconium has been applied successfully for the handling of acidic salt solutions at high temperatures. One should, as with titanium, be careful of the danger of self-ignition (pyrophoric reactions) of zirconium in some specific, extremely oxidizing media. This again includes red fuming nitric acid.

Pyrophoric reactions of tantalum are not known. Tantalum is practically universally resistant in all acidic media up to the highest concentrations. Exceptions are alkali solutions.

The special metals all have in common that they are not resistant to hydrofluoric acid and fluoride-containing solutions (also at low fluoride concentrations). All special metals are also vulnerable to nascent hydrogen (hydrogen embrittlement). In the less resistant materials, this can also be brought about by galvanic elements, especially in acidic media.

6.6
Resistance Properties of Plastics

The resistance of plastics in inorganic acids and bases is usually good. In some cases, solvents in the media used for the processes can lead to problems; even water can cause difficulties with synthetic resins. The resistance properties of a selection of plastics are given in Tab. 6.5; these properties are not universally applicable, but can be drastically limited by factors such as temperature and mechanical stress. Against this background, this table should rather be considered as a negative resistance list. A plastic indicated to have the required resistance should only be considered a candidate for corrosion experiments imitating the conditions in practice.

Tab. 6.5 General indication of resistance of plastics to various media

	PVC	PE-HD	PP	PVDF	FEP	PFA	PTFE	GF-UP	GF-EP	PF	EPDM	FPM
Water, cold	+	+	+	+	+	+	+	+	+	+	+	+
Water, hot	○	+	+	+	+	+	+	○	○	○	+	+
Weak acids	+	+	+	+	+	+	+	○	+	○	+	+
Strong acids	+	+	○	+	+	+	+	−	−	−	+	+
Oxidizing acids	○	−	−	○	+	+	+	−	−	−	−	+
Hydrofluoric acid	○	○	○	+	+	+	+	−	○	−	−	+
Weakly alkaline solutions	+	+	+	+	+	+	+	−	+	+	+	+
Strongly alkaline solutions	+	+	+	○	+	+	+	−	○	−	+	+
Inorganic salt solutions	+	+	+	+	+	+	+	+	+	+	+	+
Aliphatic hydrocarbons	+	+	+	+	+	+	+	+	+	+	−	+
Chlorinated hydrocarbons	−	○	−	+	+	+	+	−	○	○	○	○
Alcohols	+	+	+	+	+	+	+	+	○	+	+	+
Esters	−	+	○	○	+	+	+	○	○	○	○/−	○/−
Ketones	−	+	○	○	+	+	+	−	○	+	○/−	○/−
Ethers	−	○	−	○	+	+	+	○	+	+	−	−
Organic acids	○	+	○	+	+	+	+	−	+	○		○
Aromatic hydrocarbons	−	○	○	+	+	+	+	+	+	○	−	○
Fuels	−	+	○	+	+	+	+	+	+	○	−	+
Fats, oils	+	+	+	+	+	+	+	+	+	+	−	+
Unsaturated chlorinated hydrocarbons	−	−	−	−	+	+	+	−	−	−	−	−

+ resistant, ○ limited resistance, − not resistant

6.7
Cladding and Lining

Cladding and lining make different plant-part designs possible; such constructions are motivated by the need to separate the functions of the base material, responsible for pressure-bearing and stability, and the overlay material, ensuring corrosion resistance.

The base material is usually un- or low-alloyed steel. For the chemically protective overlay, on the other hand, almost any of the corrosion-resistant metals, glass (with enameling), or plastics may be used.

A special feature of, and very important in, chemical technology is chemical enamel, which has the almost unlimited chemical resistance of glass in acids and salt solutions. The corrosion properties of glass and therefore also that of enamel is determined by the solubility of SiO_2 in the medium. This is very low in acids, but in alkali and alkaline solutions, it can be rather high. This explains the relatively low resistance of enamel in strongly alkaline solutions, especially at high temperatures. An exclusion criterion for the use of enamel is the presence of fluorine ions, even in very low concentrations (e.g., 10 ppm), in the operating medium; small quantities of SiO_2 in the medium can, on the other hand, sometimes substantially reduce the aggressiveness of the medium towards enamel.

Chemical enamel is applied in multiple layers (generally between five and seven firings). It consists of two enamel prime layers and three to five enamel cover layers. The resistance of the enamel is determined by the enamel cover.

The enamel is fired at temperatures between 950 °C (enamel prime layer) and 800 °C (last enamel cover layer). After cooling to room temperature, the glass is set under compressive residual stresses, because of the larger thermal expansion coefficient of the base material. This lends the glass some thermal and mechanical resilience. As a result, a temperature difference of approximately 120 °C (e.g., when a cold product is loaded into a hot vessel) is a shock well tolerated by the glass. Enamel is, however, not resistant against mechanical blows. Also, when enameled vessels and piping are assembled, preventive measures should be taken to prevent damage to the enamel when the flange screws are tightened.

The substantial hardness of enamel provides good protection against frictional wear and tear. The smooth surface facilitates cleaning and is of great advantage in applications where products polymerize at the vessel sides. An additional advantage is that the corrosion products of enamel (if they form) are biologically and medically safe.

The coatings that can also be considered if plastics are used are, apart from phenol–formaldehyde stove enamelling, sintered-in coatings of various thermoplastics and fluoroplastics. More often, though, claddings in the form of linings are used, which have the advantage over coatings that they are thicker. This is important not only because damage can then be tolerated to a greater depth, but also because it functions as a diffusion barrier. Especially at negative temperature gradients across the walls (e.g., product is hotter than the wall), corrosion-critical media components (e.g., chloride, humidity) can diffuse through the plastic layer and condense onto the pressure-bearing steel constructions (piping, casing). The coating can be lifted (bubbles) and the steel attacked severely. In some cases, the insulation around the operating unit can help prevent this.

Loose claddings of plastic or, also, metallic materials have the advantage that leak monitoring of the cladding can be carried out by means of bore holes in the pressure-bearing wall. Stability at underpressure can, however, not be guaranteed. There is an increased danger of cracking in plant components that undergo great changes in mechanical or thermal stress, for example, at loose cladding around nozzles, since large strain concentrations are found in such areas.

In such cases, metallic cladding or, in the case of plastic liners, adhered linings are preferable. Cladding of metallic parts can be by weld-overlay cladding, explosive cladding, or roll cladding. Roll cladding has the advantage that large plates can be manufactured this way, so that fewer weld joints are required in the construction part. This method can, unfortunately, not be used for all material combinations. This is even more so for weld-cladding. With explosive cladding, though, almost all metallic materials can be bound tightly together.

6.8
The Use of Corrosion Experiments for Choosing Materials

In the previous sections of this chapter, the thermal, mechanical, and, especially, the corrosion properties of highly corrosion-resistant materials were discussed. The corrosion resistance as deciding factor for choosing a material is especially difficult to be quantified in a way that directly predicts its applicability in complex operating mixtures. A large variety of factors, such as pH, flow rate, temperature, anionic and cationic content of the solution, and, importantly, the corrosion potential of the material as determined by the redox potential of the solution and the corrosion resistance of the material, but also questions such as whether heat transfer is possible, play decisive roles. Because of this, lists regarding the resistance can only be regarded as estimations; materials appearing suitable should be regarded as candidates, at most, for the necessary corrosion experiments which should simulate the operating conditions as closely as possible.

Corrosion tests in operating- or pilot plants are preferable to simple immersion tests in the laboratory. The requirements of standards like DIN 50905 (DIN, German industrial standards) should be followed when these corrosion tests are carried out.

Fig. 6.7 shows some sample types usually used for such tests. The samples should always contain a welded joint. The choice of sample should fundamentally take the potential types of damage to the material, where the medium has a vital influence, into account.

Stress-corrosion cracking is a very critical type of corrosion, and is worth further emphasis. Tab. 6.6 shows a selection of medium-material combinations for which stress-cracking corrosion is a potential problem.

To take care of such situations, samples should be prepared as bending-test samples under stress, or, in the case of samples of corrosion-resistant steels or nickel-based alloys, with the one side coarsely ground. The coarse grinding, carried out with a size-80 disk, induces tension into the surface. A prerequisite for the use of such samples, though, is that no substantial general corrosion occurs. Also, if meaningful results are to be obtained with these simple samples, the scaling-down of the samples should be within limits and the temperatures should not be too high.

Fig. 6.7 Probes for corrosion experiments

CR 7S 179-4/16

Tab. 6.6 Examples of systems susceptible to stress-cracking corrosion

Alloy systems

C steels	Austenitic steels	Al	Cu	Ni
H_2O/NO_3^-	H_2O/Cl^-			H_2O/HF
H_2O/OH^-	H_2O/OH^-	$H_2O//Br^-$	H_2O/NH_3	H_2SiF_6
H_2O/CN^-				
H_2O/PO_4^{3-}	H_2O		H_2O/OH^-	H_2O/Cl^-
H_2O/SO_4^{2-}			$H_2O + -$ amine	H_2O/OH^-
H_2O/CO_2	H_2O/Br^-	H_2O/I^-	$H_2O + -$ citrate	Polythionic acid
$H_2O/CO/CO_2$			$H_2O + -$ tartrate	Chronic acid
H_2O/H_2S	$H_2O/H^+/SO_4^-$	N_2O_4	H_2O/NO_3^-	Acetic acid
$H_2O/FeCl_3$			H_2O/NO^-	NaOH melt
Raw methanol	Polythionic acids	HNO_3	H_2O/SO_4^{2-}	(or KOH)
NH_3			$/H_2O/SO_3^{2-}$	Vapor
(liquid and vapor)	H_2O/H_2S	Organic liquids	H_2O/S^{2-}	
$H_2O/HCO_3^-/CO_3^{2-}$	H_2O/H_2SO_3		H_2O/F^-	Organic liquids
	H_2O/NH_4OH	Moist air	Vapor	

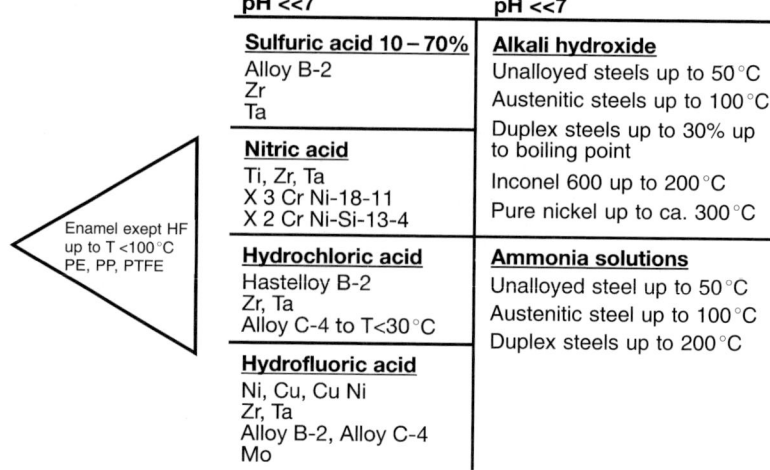

Fig. 6.8 Materials for various aggressive chemically corrosive media

18/10 CrNi steels	–	halide-free salt solutions, high product-purity requirements
17/12/2 CrNiMo steels	–	neutral to alkaline, slightly chloride-containing solutions, organic acids
22/5/3 CrNiMo steels Duplex steels	–	chloride-containing media with pH > 4 ammonia solution T > 100°C
Special steels with 20–33% Cr and/or 5–7%	–	acide, chloride-containing media with high oxidation potential, sulfuric acid

Fig. 6.9 Main application areas for corrosion-resistant steels

The extent of the corrosion is gravimetrically determined (mass before and after experiment). An optical and, if required, metallographic examination is additionally necessary for determining whether local corrosion (e. g., pitting- and stress-corrosion cracking, selective corrosion of the welded joint) has occurred, and whether the characteristics of the attack is connected to the existence of specific structures.

Metallic samples should be fitted in a way that electrically isolates them from one another, and also from the walls of the apparatus.

Figures 6.8 and 6.9 give guidelines on media for which specific materials are candidates for further tests and application areas in which the various materials have been successfully used.

6.9
References

[6.1] DEGENKOLBE, J., HANECKE, M., Rohre, *Rohrleitungen, Rohrleitungstransport* **1978**, *17*, 514–520.

[6.2] GRÄFEN, H. *Chem.-Ing.-Tech.* **1988**, *60(9)*, 662–671.

7
Process Instrumentation, Control Equipment, and Process Analysis Measurement Technology

Modern process control engineering (PCE) has become an extensive branch of engineering, as economical production is often only possible by automated manufacturing. It has found application in virtually all areas of manufacturing industry. A long tradition lies behind process control engineering itself; stations of its development over time are described in references [7.1] and [7.2]. The increasing competition of the international market demands constant improvement in the efficiency of the production processes. This is generally associated with increased demands for an automation concept and the systematic description of automation tasks. On the basis of the data generated by the process and with human contribution, suitable measures should be taken to ensure that the process runs and is documented in the required fashion (see Chapter 8). For the modern instrumentation of chemistry plants, the interdisciplinary work between chemistry and process engineering is therefore crucial. This is especially the case in the areas of quality control [7.3, 7.5], plant safety, and environmental protection (see reference [7.6] and Chapter 10).

The use of process automation systems (PAS) opens several possibilities for collecting the available information on the state of the process, including material and energy streams. Field-level process data, in a suitably prepared and concentrated form, is gaining increased importance, also from logistical and economic points of view. The functions at the field- and process control levels (Tab. 7.1) are standardized to a large extent, and are more often realized than the functions at the operating- and production control levels (see Chapter 8).

According to the hierarchical model of levels of production (Fig. 7.1), all the functions that require intensive data exchange and make use of similar equipment technology or software are grouped together.

The structuring of process control engineering into a hierarchical model of levels with basically self-sufficient function levels is of advantage in the case of malfunctions, the employment of adequately trained staff, and time structuring for realizing the individual levels. On the process control level, the individual apparatus (simple controller, programmer, recorder, etc.) are available, on the one hand, and, on the other hand, programmable logic controller (PLC) or process automation systems (PAS), which combine several of these self-sufficient functions, are present.

Tab. 7.1 Functions at the field- and process control levels

Functions at the field level	Functions at the process control level
Measurement	Logic control, open loop control
Actuating	Closed loop control
Display	Counting
Operating	Monitoring
Alarm	Analysis
Data input	Recording
Data output	Display
	Operating
	Data collection
	Data input
	Data output
	Data processing
	Handling of malfunctions
	Sequence control
	Recipe handling and control

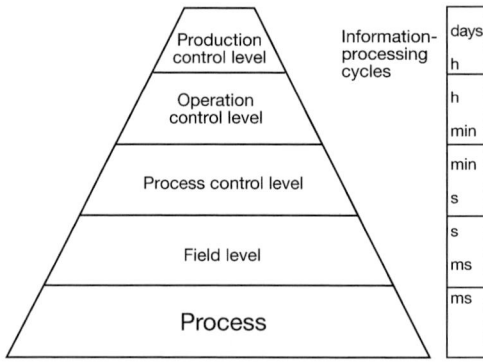

Fig. 7.1 Information flow in the pyramid of the model of levels of production

The individual apparatus at the field level can be connected to the control level through standardized interfaces. Electrical point-to-point connections are still almost exclusively used. This technology using pneumatic and electrical standardized analog signals, may, however, be pushed out by standardized, digital field-bus concepts in the near future. In this regard, the developments on the field level are clearly limping behind those of digital control technology.

Establishing the primary measurement-technology task is an important part of the total automation task. This is where the actual connection between process- and information processing is defined. From the process itself and existing environmental- and production conditions, requirements of the field equipment technology develop regarding material choice (Chapter 6), mechanical (Chapter 5) and electrical connections, stability, and circuitry (explosion protection, safety facilities, etc.).

Fig. 7.2 Process for choosing field equipment

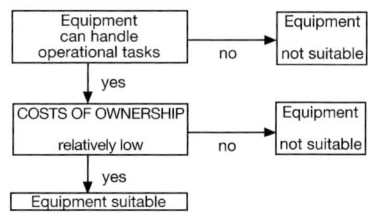

Generally, the criteria shown in Fig. 7.2 should be taken into account when equipment is chosen at the field and control levels.

Especially at the field level, the expenses associated with running a piece of field equipment may substantially exceed the acquisition price of that same equipment. Despite this, the decision for buying equipment is often based on the acquisition cost only. The buying decision-making should rather be based on what can be summarized under "cost of ownership" [7.7], namely, all the equipment-related expenses, acquisitions costs, choosing costs, testing costs, standardization costs, expenses for maintenance and repair, training costs, and the costs for planning, assembly, and start-up.

In Section 7.1, we will introduce the subject of sensors and also process analysis measuring equipment, actuators, and wiring principles. The advantages and disadvantages of specific individual field equipment will be discussed briefly, without their application in a multiproduct plant being referred to directly in each individual case (for more on this subject, see also references [7.8, 7.9]).

The multiproduct-plant concepts described in Chapters 1 and 3 are subject to different demands regarding structural flexibility and flexibility of capacity, the product spectrum, available space, and marginal operating conditions. How this is reflected in the technology of the process control engineering equipment will be discussed in Sections 7.2 and 7.4.

In Section 7.2, the communication of sensors and actuators with the process control level and process connections in multiproduct plants will be described.

Next, the criteria used for choosing equipment will be outlined. We will finish off by briefly highlighting aspects regarding the automation of multiproduct plants and plant safety with process control engineering means.

7.1
Sensors and Actuators

7.1.1
Flow, Quantity

Flow measurements can be divided into volume-flow measurements and mass-flow measurements. The equipment for this produce an electrical signal that is directly connected to the volume or mass flow of the gases or liquids. This should preferably be a current signal of 0/4 to 20 mA, or, for more precise applications or total volume measurement, be a frequency signal.

For dosing, total volume measurement, and so forth, the volume or mass flow accumulated over a time period is to be measured. If the total mass is the most important aspect of the measurement, equipment based on balance technology is preferable, since these give a signal directly proportional to the weight.

Recommended Equipment

Volume flow meter for liquids, based on the magnetic induction method (magnetic flow transmitter)

A minimum conductivity of approximately $10\,\mu S\,cm^{-1}$ is required. This type of meter is usually reliable and accurate; the error in measurement is typically approximately 1% of the measured value. If the measured values lie below the minimum volume flow specific to that apparatus, larger errors can be reckoned with.

Mass Flow Meter, based on the Coriolis Method

These meters are designed for liquids, but may also work with compressed gases. They have been used successfully even where multiphase flows are present. The Coriolis mass flow meters are generally reliable and accurate, with errors typically less that 0.5% of the measured value. Several apparatus of this type also have an additional density measurement function.

Volume Flow Meter for Gases

The volume flow of gases can be determined with orifice plates, Venturi tubes, plummets, turbine-wheel counters, thermal mass flow meters, or Coriolis mass flow meters (at suitable gas density). The display, however, is strongly influenced by the operating conditions (density, pressure, temperature, heat capacity, viscosity, etc.). Especially with the often used orifice and plummet flow meters, the measurements depend strongly on the material density and viscosity, so that the meters need to be calibrated for every operating state (work point). This greatly limits their application as universal measuring devices.

Weighing Cells with Strain Gauges

With this type of weighing cell, the deflection is measured while an elastic element is stretched within its Hooke range. For the sake of accuracy, signal analysis proceeds almost exclusively digitally nowadays, as analog amplifiers are at most capable of an accuracy of ±0.1%. Weighing cells with strain gauges are used in process balances, filling units, and almost all industrial balances. The measuring range is from 1 kg to 1000 t.

Weighing Cells with Oscillating Strings
The weighing cell functions by measuring the resonance frequency of a metal string, in which an alternating current flows, in a magnetic field. It is mainly used for differential dosing balances (measuring range 3 kg to 1 t).

7.1.2
Pressure, Differential Pressure

The equipment is used for determining pressure (overpressure, p_e; absolute pressure, p_{abs}; differential pressure, Δp), either to be displayed directly in location (manometer) or to be converted into a proportional electrical signal (0/4–20 mA, 0/1–5 V, 0/2–10 V) (pressure transducer).

Recommended Equipment
The techniques available for measuring the absolute- and relative- or differential pressures are described below.

Manometer
Manometers operate mechanically. The deviation of the needle can be intercepted, without direct contact being made, at defined detection values (limit transmitter) or continuously (remote pick-up). The accuracy of the signal produced by the manometer with remote pick-up is typically within 1.0% of the upper limit of the measuring range.

Pressure Transducer
Pressure transducers operate electrically. The pressure-dependent deformation of a membrane flush with the front leads to a proportional change in an electrical value (capacity, resistance: strain gauge, piezo-resistant elements). The accuracy of the measurement is usually around 0.2% of the upper limit of the measuring range.

Single-Range Transducer
The signal produced is limited to a given pressure range (e.g., 0–16 bar). The advantages of this device are its small size and low price.

Multiple-Range Transducer
Within certain limits, the signal produced can be assigned to a freely chosen pressure range. With an offset potentiometer, the lower limit and range (upper limit – lower limit) of the measurement can be set. The measurement range can typically be set in the ratio of 1:6. The pressure transducer must be calibrated with a pressure standard at a technical workshop. The advantages of this apparatus are its robustness and stability to overload of Δp transducers. The disadvantage of this device is its large size or mass.

Smart Multiple-Range Transducer

The smart multiple-range transducer functions digitally; the lower limit and range of measurement can be configured. The characteristic lines of the measuring elements are stored in the device itself, eliminating the need for calibration, unless the characteristic line itself has been changed by, for example, too high overloads.

The measuring range can, typically, be set in the ratio 1:20. Configuration is done via a PC or a so-called hand-held terminal, whereby the communication takes place in the same line used for the electrical signal (overlay of 4–20 mA signal and digital signal). For this, special circuitry is necessary for further signal processing. The advantage of this equipment is that the measurement range can be set simply on location. Its disadvantage is that additional aids for configuration are required, which can differ from supplier to supplier.

Local Pressure Indicators

As local pressure indicator without any further process control engineering functions (display, control, limiting value generator, etc.), the most economical device is the manometer.

7.1.3
Temperature

Measuring resistances such as Pt 100 and thermoelements are preferentially used for measuring temperatures from −200 °C to 1000 °C. The sensors are pre-manufactured measuring inserts with diameters of 3 mm or 6 mm. The actual temperature-sensitive part is at the tip. Mantle thermometers can be fitted directly via screw fittings on location. For maintenance purposes, the inside of the apparatus and piping should be separated from the surroundings by a thermometer protective tube. The measuring device is pushed into this protective tube. A measuring amplifier transforms the temperature signal into an electrical signal. For installation in the field, head transducers are used and normal transducers are used for installation in the control room.

Recommended Equipment

Resistance Thermometer

Resistance thermometer Pt 100 for the temperature range −200 °C to ca. 400 °C (accuracy class A $<\pm 1$ °C, total measurement accuracy $<\pm 1\%$ of the measurement range).

Pt 100 resistance thermometer inserts, diameter 3 or 6 mm.

Preferred use of single Pt 100 in four-wire connection.

Thermoelements

Thermoelements for the temperature range $1\,°C$ to $1000\,°C$ (accuracy class for the sensor up to $400\,°C$ is $\pm 3\,°C$, from $400\,°C$ to $1000\,°C$, it is $\pm 7.5\,°C$, total measurement accuracy ca. $\pm 1.5\%$ of the measurement range).

Because the wall thickness of the protective tube of thermoelements is kept to a minimum for given materials and close fitting of the measuring insert, fast and very accurate measurements are possible. It is important that at least the temperature-sensitive part is completely surrounded by the medium. Heat-conducting pastes for improving the measuring performance are generally not used.

7.1.4
Level, Interface

The purpose of the measurement is the detection of the state of fullness of a container (liquid or bulk material), limiting levels (high, low), or an interface, and the conversion into a proportional electrical signal.

Recommended Procedure

Electrical conductivity $\sigma \approx 0$: $\rightarrow$ Hydrostatic measuring method (e.g., smart differential pressure transducer)

The measuring principle is based on the determination of the pressure that a liquid column of height h and density ρ applies to the bottom of a container. Because the pressure, p_h, is linearly dependent on the density, ρ, errors result from density changes (e.g., changes in product composition or density variations resulting from temperature changes of the product; measurement errors resulting from temperature changes are typically 1% per 10 K). A measurement is, however, always there, in contrast to mechanical methods, such as floating probes for measuring levels ("drowning" of the floater when the density is lower than the minimum density, ρ_{min}). Problems associated with hydrostatic methods are:

- Condensed medium, collecting in the minus line of the differential pressure measurement lead to measurement errors (when possible, solved by flushing with inert gas)
- Precipitation of solids can block the process connection of the plus line
- At high pressures, a change in operating pressure can lead to a measurement error

Electrical conductivity $\sigma \geq 1\,\mu S\,cm^{-1}$: $\rightarrow$ Conductiometric method

The measuring device (probe) consists of a metal rod, which is inserted vertically from the top of the container and dips into the liquid to be measured. The difference in voltage between the container wall (reference electrode) and the top end of the rod is proportional to the level h of the container (potentiometric principle). For the measurement to function properly, the liquid should have a conductivity, σ, of at least $1\,\mu S\,cm^{-1}$. The measured value is unaffected by changes in the electrical conductivity in the container. This conductiometric method cannot be used with media that tend to form films on the walls and probes.

Level measurement by the admittance principle

- For low electrical conductivities, down to $\sigma = 0.05\,\mu S\,cm^{-1}$
- For operating temperatures up to 200 °C
- For film-building media (films on the probe are electronically recognized and are compensated for)

Because it is very expensive, this probe is only used for special cases.

Optoelectronic Limit Indicator

Operating principle: The reflection of infrared light in a glass rod is measured. This measuring principle is well suited for transparent, non-adhesive liquids. It is unsuited for optically dispersive liquids (e.g., milk).

Thermal Limit Indicator

Operating principle: For this measurement, a resistance at the end of the probe is heated by a defined current. This method can only be used in the temperature range −25 to +80 °C.

Mechanical Limit Indicator with Vibrating Fork

Operating principle: The change in the resonating frequency of a piezoelectrically stimulated tuning fork is measured. This method is suitable for viscosities up to 2000 mPa. Deposits on the tuning fork can be problematic.

7.1.5
Process Analysis Measurement Technology

The purpose of process analysis measurement technology is to continuously measure the concentration of a substance or several substances (liquid or gas) in product lines or containers.

Process analysis measurement technology is, in principle, subject to the same requirements placed on normal plant measurement technology: the measuring equipment must be reliable, robust, and easy to use. It, nevertheless, plays an especially important role within the general process engineering. This is mainly because of the different problems faced by a process analysis measurement device. Whereas pressure, temperature, and flow-rate sensors directly determine physical quantities and make these available in the form of a display, process analysis instruments use chemical and physical effects for determining the concentration of one or more components. This also explains the potential dependence of the measured concentration on interference from accompanying components, state parameters, or environmental influences, which affect the basic measuring effects (cross-sensitivity).

An absolute requirement for reliable concentration values from an analytical measurement in a plant is that the process engineering data should be available in an as complete as possible form in advance. Apart from data on the states of the materials to be measured, such as pressure, temperature, density, melting point, and solid content, information about explosion protection, usable materials, and required 90%-time of the measuring value is also necessary. If the possibility of false measured values due to cross-sensitivity of the measuring method towards accompanying components in the medium is to be excluded from the start, the exact composition of the sample should be known, including the type of substances as well as their possible concentration ranges.

Malfunction safeguarding is an extremely important factor for ensuring continuous analysis. The analysis device therefore needs to be substantially independent of external influences such as temperature, vibration, dirt, or the influence of corrosive media.

The accuracy of analysis instruments in plants is generally determined by calibration, which, in turn, is determined by laboratory equipment. The plant analysis can therefore not be more accurate than the laboratory analysis. In contrast, the reproducibility over a longer time period is generally regarded to be better than with laboratory equipment. This does, of course, include good drift behavior.

Process analysis, even when discontinuous, as in chromatography, is always faster than laboratory analysis, owing to the automation of sampling, sample transport, measurement, and analysis. In especially demanding situations, the plant needs to be optimized regarding the parameters determining the rate.

The maintenance of process analysis measurement equipment is usually carried out by plant personnel without in-depth technical training. For this reason, factors such as simple calibration, being low-maintenance, the possibility of easy and fast repairs on location, and simple, straightforward operation are crucial. These requirements are easily fulfilled by modern equipment technology, which generally offers automation together with easy-to-operate equipment.

The following compilation presents an overview of the most important measuring effects used as industrial analytical tools.

The measuring principles can be divided into two groups. The first group consists of measuring devices that make use of a directly generated physical quantity, such as the absorption of light of a characteristic wavelength, for determining the concentration of a substance. To this group belongs:

- Optical absorption
- Paramagnetism
- Heat conduction
- Electrical conductivity
- Dielectrical constants
- Density
- Adsorption

The second group consists of analytical equipment where the actual measured value is produced by an auxiliary chemical or physical reaction. The following measuring principles are included:

- Calorimetry
- Electrical conductivity after a reaction
- Electrochemical effects
- Chemiluminescence
- Radiation absorption after a preliminary reaction

How these methods work and their general application possibilities are described in detail in the specialized technical literature [7.10].

Installation of a Process Analysis Device

After the measuring task has been defined, a suitable analytical method and, subsequently, suitable equipment need to be chosen. It often happens that the specific measuring task cannot be solved without conditioning of the material to be measured.

Whereas other measuring devices are usually built into the container or piping (in-line), this is often, in principle, not possible or useful with process analysis devices.

In-Line Measurement

The advantages and disadvantages of in-line measurements, where the measurement probe is placed directly in a product line, apparatus, or container, are set out below.

Advantages	Disadvantages
Assembly	Maintenance
Sampling unnecessary	No sample preparation possible
Sample transport not required	Calibration difficult
Fast, immediate indication	
No sample disposal	

Since in-line measurement proceeds without sample-taking, transport, preparation, and disposal, the effort of assembly is relatively minor.

This advantage is offset by the significantly greater maintenance effort generally required. Maintenance and repair work that necessitate removal of the measuring device is usually not possible during operation, and are often additionally hindered by local conditions or explosion-protection conditions.

A substantial advantage of in-line measurement is that the measurement is displayed fast and immediately, since there are no blank time gaps due to the transport of a sample.

Fig. 7.3 Various steps of an on-line analysis procedure

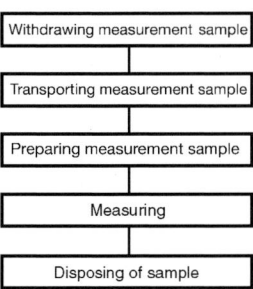

The following measurement principles are most often used in-line:

- pH values
- Conductivity
- Redox potentials
- Moisture content
- Density
- Scattered light.

On-Line Measurement

The majority of the process analysis measurement equipment is equipped with on-line technology. The measurement proceeds in a bypass flow, which can be disconnected for maintenance of calibration.

An on-line measurement comprises the individual steps shown in Fig. 7.3.

All the steps need to be carefully and thoroughly considered to ensure a representative sample and to exclude possible influences on the measurement effect, accuracy, and time. The often required sample preparation usually means a change in parameters such as temperature, pressure, and solid content and the exclusion of interfering components; depending on the complexity, this may result in a small process engineering unit dedicated to this function.

7.1.6
Signal-Processing Equipment

Signal-processing equipment is of decisive importance in process control engineering. It provides the connection to the field level, on the one hand, and represents the interface with people, on the other hand. Individual signal-processing devices are dotted line- and chart recorders, displays, controllers, signal converters, and limit transmitters. The large variety of equipment is a result of the numerous requirements (registration, display, control, etc.) and the many signal forms (4–20 mA, 0.2–1 bar, etc.).

Dotted line recorders, chart recorders, and controllers are installed in conventionally equipped control rooms or local control stations. If the plant is equipped with a process automation system, these individual devices are used only in exceptional cases.

Signal converters, on the one hand, interconvert different forms of signals and, on the other hand, separate different electrical circuits of a signal route. With modern signal converters, it is generally possible to monitor preset limiting values and to set off the relevant switching if necessary. Signal converters are not necessary where no potential separation or explosion protection is required; it is then usually better to connect the field equipment directly to displays, controllers, and process automation systems. In multiproduct plants, the configuration and parametrization should preferably be saved with the use of software in nonvolatile form on the converter. The total transfer error should be less than 0.5% of the measured value, and be insensitive to fluctuations in the auxiliary power.

7.1.7
Control Valves

The purpose of using control valves in chemical processes is to influence material flow, by allowing more or less flow through the opening intersections. Apart from control valves operating on the simple *open/closed* basis (e.g., ball valves), gradual changes in the material flow are also required for control purposes. A control valve consists of an actuator and the valve. The most commonly used valves are globe valves, ball valves, and butterfly valves. These flow-control devices are supplemented by gate valves, vibrating feeders, frequency-controlled pumps, and so forth. The control valves for continuous control used most often in the chemical industry is the globe valve with pneumatic actuator.

Recommendations for Use

Ball Valves
The low-price three-way ball valves should only be used for simple applications as flange and welded valves. Alternatively, two-part ball valves with floating ball design or two-part ball valves in heavy design from DN 80 upwards should be used for more demanding situations in terms of pressure difference, temperature, and use in critical media. PFA (perfluoroalkoxycopolymer) can be recommended for the lining of coated ball valves, because, not only is its chemical resistance comparable to that of PTFE (polytetrafluoroethylene), it is also more diffusion-resistant and inexpensive than PTFE.

Where special corrosive attack is an issue, other materials, such as Hastelloy steel, tantalum, titanium, and so forth, can also be used.

The tightness of ball valves in the closed position is very good, and in the open position almost the full line diameter is available.

Butterfly Valves
The preferred use of butterfly valves is for controlling the flow through lines of larger diameters (e.g., DN 80, 100, 150, 200, etc.), since, in these applications, they are more economical than ball valves and globe valves. The control ability of butterfly valves is rather poor.

Globe Valves

In globe valves, a valve spindle connected to a cone moves the cone relative to the valve seat, which is permanently joined to the valve body. Since globe valves are especially used for continuous control, for dimensioning the required k_v values should be calculated from the given flow- and material data. For dimensioning, the possibility of further flow resistances installed before or after the control valve should be considered. Especially in multiproduct plants, the valve trim may need to be exchanged more often. For the choice of material, the points discussed under ball valves are also valid for globe valves.

To ensure the required tightness towards the outside, it is necessary to seal especially the valve. For this, elastic membranes, bellows, or packing glands are usually used. The packing gland is exposed to the product on the inside, so that suitable materials also need to be used for this.

Actuators

In chemical process applications, actuators are operated almost exclusively pneumatically. Spring-loaded or non-spring-loaded drives provide the lifting- or turning motion, with as little delay as possible. This is important for the dynamic behavior of control loops during startup and shutdown. With continuous controllers, an associated electrical signal is assigned to the corresponding turning or lifting. Imprecise settings resulting from large frictional forces can be avoided by the use of positioners with proportional action.

7.2
Communication between the Field Level and the Process-Control Level – Process Connections in Multiproduct Plants

There is a very wide range of technology associated with individual sensors and actuators at the field level. The individual devices are usually in direct contact with the products and, for this reason, firstly need to be linked to the multiproduct plant by suitable, preferably standard process connections. The connection technologies used for the tubing and apparatus in a multiproduct plant with regard to the choice of materials and connection types are essentially determined by the specific process engineering requirements [see Chapter 5; e.g., sandwich-type construction for installation between flanges or threads of nonmetric screws, such as G 1/2, NPT 1/2″ (NPT, U.S. standard taper pipe thread) or metric threads such as M 20×1.5]. In contrast to chemical monoplants, clearly defined data on volume flow, pressure, viscosity, corrosivity, and so forth are usually not available.

The expenses associated with repairs, plant modifications, fitting or removal of sensors when the measuring range needs to be adjusted, and commissioning should therefore be expected to be higher in multiproduct plants than in monoplants. The use of screw fittings and flanges is therefore recommended in multiproduct plants.

The interface to the process control level is also currently almost exclusively through point-to-point connections [7.1, 7.11]. Standardized electrical signals are used (0/4–20 mA, 0/1–5 V, 0/2–10 V). Two-wire technology is applied (but also four-wire technology for temperature measurements). Signal transmission proceeds through transducers, connection boxes in the field, fixed cables, and terminal blocks with jumper wiring in the control room up to the process control level.

If one considers the different types of multiproduct plants, it is clear that especially modular multiproduct plants and pipeless plants will have difficulty with regard to their structural flexibility with fixed electrical wiring structures.

In modular multiproduct plants, a mobile plant section (see Chapter 8) only occupies a fixed location relative to a permanently placed reactor within the framework of a specific campaign. In pipeless plants, in contrast, only the reactor is mobile, and, according to the process sequence, is connected at various permanently located stations. In both the above cases, the mobile plant sections can be regarded as units that need to be flexibly attached to the electrical or pneumatic power supply.

In this regard, plug concepts are recommended. Electrical connections via plugs in non-explosion-hazardous areas should comply with the NAMUR recommendation NE 28 (NAMUR, Normenarbeitsgemeinschaft für Mess- und Regelungstechnik in der Chemischen Industrie: the German Association for Standardization in Measurement- and Control Technology in the Chemical Industry).

To what extent these units can or should be self-sufficient with regard to their measurement and control facilities will be discussed in Section 7.4.1.

Standard multiproduct plants and multiproduct plants with pipeline manifolds have no problems in this regard.

In principle, for example, pumps of these multiproduct plants could be supplied with plug connections if it appears as if they may have to be exchanged often because of crystallizing or corrosive media. This is, however, generally only economically viable if undertaken during construction of a new plant, as upgrading afterwards can be costly.

Currently, programmable digital interface protocols (e.g., Modbus, manufacturer-specific protocols) are seldom used for providing connections at the field level. This is often only used for linking balance signals or for connecting programmable logic controllers. A uniform, standardized field-bus concept is not available yet.

In the current transition phase (see reference [7.13], Fig. 7.4), intelligent sensors, whose measurement range can be adjusted by a special protocol (e.g., HART protocol) through a PC or hand-held terminal (see Section 7.2) while it is still fitted, are increasingly used. This proceeds through the same cables used for the standardized electrical signals. Measurement range conversions of 1:20 for pressure measurements and 1:50 for mass flow measurements are possible.

Sensors that can be configured by use of the HART protocol (HART, highway addressable remote transducer) should be used in multiproduct plants built these days, because with these sensors a large part of the measurement range can be covered without the need for sensor exchange.

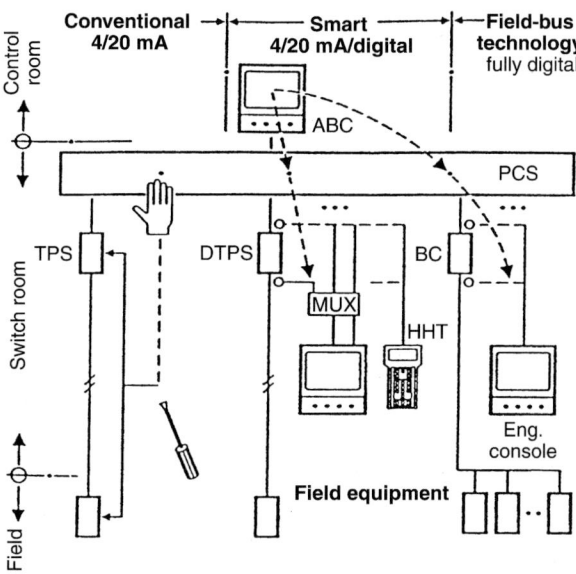

Fig. 7.4 Migration of conventional/analog field-equipment technology via "intelligent" analog field equipment with internal microprocessor technology (smart technology) to field-bus technology (TPS/DTPS, transducer power supply; BC, bus coupler; HHT, hand-held terminal; MUX, multiplexer)

Only with the field-bus concept is a comprehensive connection of measurement and control devices with pure digital technology reached. It will be associated with expanded measuring functions in the collection of the measured value (transformation of technological process quantities, amplification and filtering of the measurement signal, automatic measurement range choice), in its own specific functions (bus coupling, self-monitoring, limiting-value monitoring, auxiliary-power monitoring), and in measured value preparation (error compensation, automatic zero-point correction, measured value conversion, characteristic-curve correction, filtering out outliers).

On the control valve side, advances can be expected in the area of control functions (control, binary control, positioning, hysteresis compensation, set-value rate limit, etc.) and data transmission functions (self-monitoring, error diagnosis, bus coupling, data checking).

Furthermore, the transfer of "intelligence" into the field will result in changes in the functions of today's process-related components (PRC). We will, however, not cover this topic in further detail.

7.3
Choosing Equipment for Multiproduct Plants

The choice of equipment for a multiproduct plant is mainly determined by its special operational requirements of flexibility of capacity and product assortment. The individual multiproduct plant concepts also generally have different characteristics (see Tab. 7.2).

The influence of structural flexibility on the choice of equipment is, in contrast, only really important for the design of automation concepts (see Section 7.4.1). The pressure range of up to 3 bar typical for multiproduct plants is possible for all the apparatus, and therefore has no determining function.

The requirement of flexibility in product assortment essentially determines the spectrum of physical–chemical properties that should be handled, such as corrosivity, tendency to crystallization (encrusting, baking), density, pressure, viscosity, electrical- and thermal conductivity, and dielectric properties.

The requirements resulting from the need for flexibility in capacity are especially those regarding measurement- and setting ranges. Tab. 7.3 roughly indicates the general influences of the different types of flexibility on the different classes of equipment.

When choosing multiproduct plant sensors, one should therefore aim for the most comprehensive chemical resistance possible (see also Chapter 6). The necessary materials such as steel/enamel, glass, Hastelloy, and PTFE/PFA coatings are usually on offer by the manufacturers.

Only in pipeless plants, where flexibility in assortment is of lower priority, is it possible to tailor the sensors and actuators optimally (and therefore generally less expensively) to the relevant product class.

The measurement range demanded from flow measurement devices, as well as the control range of the control valve devices, is a result of the required flexibility in capacity. The control valves must therefore allow easy exchange of valve trims. In special cases, it is, for example, possible to acquire the desired flexibility through a parallel construction of flow measuring/control valves with additional measurement/control ranges. Essential, but realizable requirements for control devices in multiproduct plants are that they should be leakproof to the outside, be

Tab. 7.2 Equipment-related flexibility requirements of multiproduct plants

Plant concept	Product-assortment flexibility	Flexibility in capacity
Standard multiproduct plant	High	Low
Modular multiproduct plant	Medium	Low
Multiproduct plant with pipeline manifolds	Medium	High
Pipeless plant	Low	High
Monoplant	Low	Low

Tab. 7.3 Influence of multiproduct-plant flexibility types on the different groups of process control engineering equipment

Flexibility type	Influence on	Flow, quantity	Pressure, differential pressure	Temperature	Level, interface	Process analysis measurement technology	Signal-processing equipment	Control valves
Product-assortment flexiblity	Range of measurement, regulation	–	○	–	–	+	–	○
	Function	+	○	○	+	+	–	○
	Material	+	+	+	+	+	–	+
Flexibility in capacity	Range of measurement, regulation	+	–	–	–	–	–	+
	Function	○	–	–	–	○	–	○
	Material	–	–	–	–	–	–	–

+ Major influence, ○ product- and technique-dependent, – minor influence.

easily rinsed, have defined flow properties, and have very little dead volume (especially under GMP requirements).

Flow measurement devices should be as independent of the materials and as insensitive to multiphase currents (gas bubbles, solid content) as possible, have short inlet and outlet distances, and result in little drop in pressure. Of great advantage, especially with crystallizing media, is good cleanability.

For the flow measurement of liquids, Coriolis mass flow meters with broad measurement ranges, configurable on location, are recommended. Disadvantages associated with these meters are sensitivity to blockages and vibrations in the plant, pressure drop, and high purchase price. Where there is no changes in the properties of the substances handled, a magnetic flow transmitter is also a reliable alternative with little drop in pressure.

The flow measurement displayed is, however, strongly dependent on the operating conditions (density, pressure, temperature, heat capacity, etc.). The often-used methods based on orifice-plate- and plummet flow measurements are especially strongly affected by the density of the material; this means that these devices need to be calibrated for every operating condition (operating point).

This strongly limits the universal applicability of these measurement devices. Despite this, they have proved themselves in individual cases, where there is little or no fluctuation in material properties, pressure, and temperature, and the substances contain no impurities. Tasks such as purging and inertizing are examples of processes that ensure that the materials and operating conditions stay constant even in multiproduct plants, and for which, at the same time, the flow measurements need not be highly accurate. Thermal mass-flow meters are very useful at almost constant heat capacities and where slight impurities are present.

Where balances are utilized, those described in Section 7.1.1 are used. For the sake of accuracy, however, the data should be analyzed digitally. For balance facilities, a precise and tension-free construction is also required.

The requirements for pressure measurement devices are, in principle, based on the manufacturing conditions. If pressure measurement should, for example, be possible in a continuous range of 5 mbar to 1.5 bar (absolute pressure) with the accuracy staying at ca. 2%, this is at the moment only possible with two complementary single measurements. In addition, models with little dead volume or front-closing membrane should be chosen.

Standard temperature measurement with Pt100 (see Section 7.1.3) usually also meets the requirements of multiproduct plants. Product baking onto the protective covering may cause longer response times.

Greater accuracy of measurement than mentioned previously is also possible at correspondingly greater costs. This can, for example, be necessary for keeping the heat balance of the reactor heating/cooling cycle or for recording column temperature profiles above 200 °C.

If (local) point measurements are required, thermoelements are used. These are, however, not very accurate.

The level- and interface-measuring devices usually only operate up to 200 °C. The electronic part of the meter should, if possible, not be located in the connecting head of the measuring sensor (probe).

If hydrostatic measurement methods are used, multirange transducers that can be configured on location are recommended. The usual measuring methods and difficulties have already been discussed in Section 7.1.4.

Where great product-assortment flexibility is required, the function of flow-, level-, and interface-measuring devices could be limited. The general requirement of multiproduct plants that no measurement errors should result from changes in the physical or material parameters can, of course, not be met in all cases. Because of the possible broad range of physico-chemical characteristics, diverse, complementary measuring principles need to be applied nowadays. A higher level of plant availability is therefore associated with higher equipment purchase, inspection, and maintenance costs.

Reduced functioning of pressure- and temperature measurements, as well as the respective control valves, mainly result from product baking onto the devices.

Functional losses in the area of flexibility in capacity occur if, despite being carefully planned, measurement ranges are exceeded and the plant availability is thereby reduced.

For the planning of process analysis measurements, the prevailing operating conditions in a multiproduct plant is of similar great importance.

Whereas flexibility in capacity is seldom of great importance, factors such as frequently changing products, unknown or insufficiently known physical parameters (density, conductivity, etc.) can put into question the use of a measurement device, or substantially increase the costs for realizing a functioning measurement

Tab. 7.4 Application limits and problems of process-analysis techniques in multiproduct plants

	Changing product composition	Changing measurement ranges	High temperatures	High pressures	Small quantities	High viscosities (only liquid measurements)	Changing physical parameters
Optical absorption	–	○	○	–	○	–	–
Paramagnetism	○	○	–	–	+		–
Density	○	+	○	○	○	○	○
Conductivity	○	+	+	+	+	○	+
pH	+	+	–	–	+	–	○
Heat conductivity	–	–	–	–	○		–
Viscosity	+	+	+	+	○	+	+
GC	–	+	○	○	+		○

– Very limited applicability, ○ appicable in special cases or with greater technical effort, + application mostly unproblematic.

unit. Different products are, for example, often associated with different measurement ranges, different matrices, or different auxiliary physical conditions and place high demands on the flexibility and dynamics of the measurement system used.

Several measurement systems are, however, optimally adapted to the measurement range and the influence of interfering components and is therefore tailored to an *individual* analytical situation. Examples of such situations are the design of separating columns of gas chromatographs and the choice of the measurement- and reference wavelength of a photometer.

The application potential for measuring several different product-stream compositions should therefore be carefully judged case by case.

An overview of the potential of the most important analysis methods for fulfilling the requirements of multiproduct plants is given in Tab. 7.4.

A more complex measurement system such as a Fourier-transform spectrometer or mass spectrometer may also be a solution; because of their greater dynamics and ability to analyze complete spectra, they have the potential to solve, under software control, several measuring tasks. These measuring systems can thereby also be incorporated into the automation program of a multiproduct plant and solve several analysis tasks by remote control.

7.4
Automation of Multiproduct Plants

7.4.1
Automation Concepts

Dotted line and chart recorders and controllers are used in conventionally equipped control rooms or local control stations. If the plant is equipped with a process automation system, these individual devices are used only in exceptional cases.

Conventional control rooms and local control stations are possible and practicable in all four types of multiproduct plant.

However, if optimizing tasks (e.g., adapting of process conditions, automated starting up and shutting down), extended analysis of disturbances/documentation/data collection, comprehensive sequence control, and recipe handling and control (see Chapter 8) are required, the functions of the individual digital devices (e.g., controller, programmer, dosing station, programmable logic controllers) should be integrated by a process automation system. Local control stations essential to the operation (e.g., startup or shutdown) can be connected in the same way. The basic facilities for this is nowadays found and mastered by all the commonly used process automation systems. Premanufactured program building blocks (Firmware) are used for handling the process control functions.

The function of a process automation system is, on the one hand, human–machine communication via display and operation components (DOC) and, on the other hand, communication with the field level via process-related components (PRC).

The display and operation components (DOC) used nowadays are usually commercially available workstations. The non-real-time operating systems, data protocols, and networks used in these are widely standardized.

The process-related components, in contrast, are usually company-specific components with real-time operating systems.

If operational requirements demand functions at the manufacturing execution level (yield calculations, statistical analysis, quality control, etc.), these only become possible or substantially simpler by data coupling to a process automation system. For this reason, the criteria for choosing process automation systems are given below.

7.4.2
Choosing Automation Systems

The use of process automation systems in multiproduct plants only makes sense if this is of advantage to the manufacture process and the product quality.

The following criteria should be used for selecting process automation systems:

Process Control Engineering Task

- Automation of chemical process
- Measurement data collection and input of measurement and analysis data, as well as display and archiving
- Logic control and set-point input for underlying self-sufficient individual components, as well as integration of the functions of the self-sufficient components

Operating Conditions

- Configuration and parametrization by process control engineering experts, only in exceptional circumstances by trained plant personnel
- Project planning by process control engineering technical unit
- Suitable for shift operation
- In systems with recipe packages: recipe development and parametrization by trained plant personnel
- Clearly defined operating- and access hierarchy.

Technical Requirements Useful for Equipment

- System solution with hard- and software from same supplier
- Capability to process large number of measuring points
- Decentralized structure of process automation system:
- Distributed functions [process-related components (PRC), as well as display and operation components (DOC)], where the process-related components represent a self-sufficient component for dealing with process-related functions (measurement and control)
- Operation and observation are possible from various points, also at remote terminals (Ex)
- Structured and modular design possible
- Standardized connection technology in hard- and software
- High system availability (redundancies possible)
- Fast, long-term-available support

Useful Functional Characteristics

Data acquisition

- Acquisition of data of different origin (sensors, etc.)
- Standardized data storage.

Data Processing
- Representing data in tabular and graphic form
- Combining data of different origins according to time context (up-to-date measurements, historical data, analyses)
- Data export in standardized format.

Reproducibility

- Defined and reproducible production parameters
- Use of precise measurement technology.

Important Issues in Project-Related Applications

- Operation and observation
- Plant management
- Sequence control
- Recipe handling and visualization
- Manual interventions of the operating personnel in processes and procedures solved according to practical situation
- Flow-chart surface
- Simple, fast configuration and parametrization
- Realization of special applications
- Standardized interfaces to higher-ranking software

We therefore recommend the timely consideration of the individual profile requirements (operation, process engineering, process control engineering, see above or reference [7.14]) by the project team, and a match between the required profile and the systems that are under consideration.

7.4.3
Process Control Engineering Structuring in Multiproduct Plants by Process Automation Systems

Normally, when multiproduct plants are automated, important individual plant sections (or process sections) are distributed according to the process-related components into automation islands, which communicate with one another through a bus system. In this way, the availability of the entire plant is enhanced. Plant sections that may be regarded as mainly part of the infrastructure can be combined in the same way. The display and operation components (DOC) (central or on location) and systems from other sections are also connected to the bus system. The individual components communicate over the bus system with one another (see Fig. 7.5).

Structuring the hardware of stationary plants, such as the standard multiproduct plant and multiproduct plants with pipeline manifolds, is not associated with any fundamental problems and is based on the structuring considerations outlined in Chapter 8. With it, all the interlocks within these multiproduct plants (including that of the manifolds) can operate through the process automation system. Safety-related facilities, in this case, must not be realized through the process automation system (alone).

More difficult to treat here are the non-location-bound plant sections of modular multiproduct plants and pipeless plants. One way in which these location-changing plant sections may be automated is if they are independently equipped

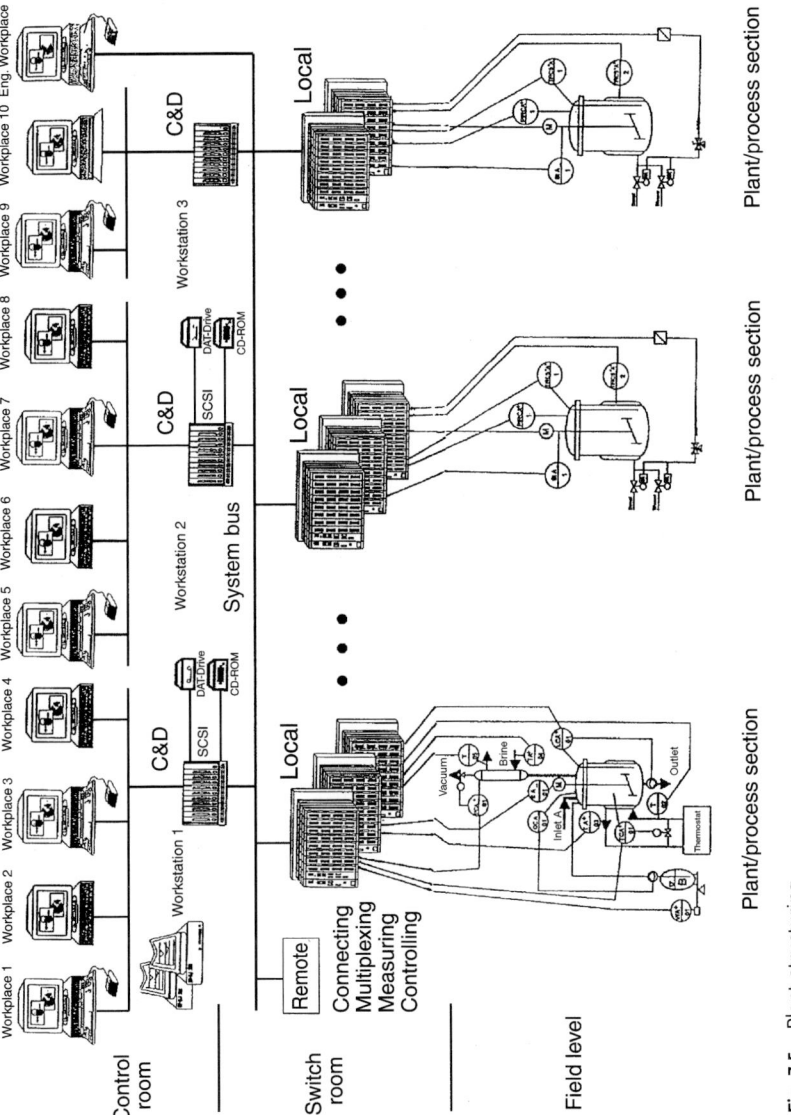

Fig. 7.5 Plant structuring

and interlocked. All process control engineering facilities (including pump motors) are then allocated to that plant section and are moved together with that unit. Interlocks may be hard wired or be served by a programmable logic controller. Then only the necessary electrical or pneumatic sockets need to be available for the power supply at the different locations. If there is no data technological connection to a higher-ranking process automation system, a "data island" is formed, which could also be equipped with "data collectors" (recorders, etc.) that move with the unit. Problems may result here from interlock concepts which have to be used from several plant sections together.

However, if one plans to have standardized digital interfaces, for example, programmable logic controllers for the movable plant sections, the allocated data can be transferred variably over designated socket installations too (with possible explosion-protection issues addressed) at the application location in a process automation system. This then opens the way again for advanced functions and interlock concepts.

Individual control units (see Chapter 8) function in a self-contained manner and differ from monoplants only by their particular selection of sensors and actuators.

7.5
Process Control Protective Devices for Plant Safety

Safe operation of a multiproduct plant can, in principle, be achieved by process engineering/equipment measures, process control measures, and organizational measures.

During the planning stage, normal operation of the plant is defined. Whether a *protective facility* is required or not is determined by a risk evaluation procedure (see Fig. 10.2). The risk determined in this way should not exceed the risk limit R_{lim}.

In Fig. 7.6, an additional catalogue of questions is given for the determination of the necessity of protective facilities within the context of the plant safety concept.

Step 1: Examination of Possible Failures and their Frequency

Technical Failure
- Breakdown of machinery and auxiliary facilities
- Power failure (electrical, pneumatic)
- Faults or failures in the process control engineering operating- and monitoring units
- Failure of seals

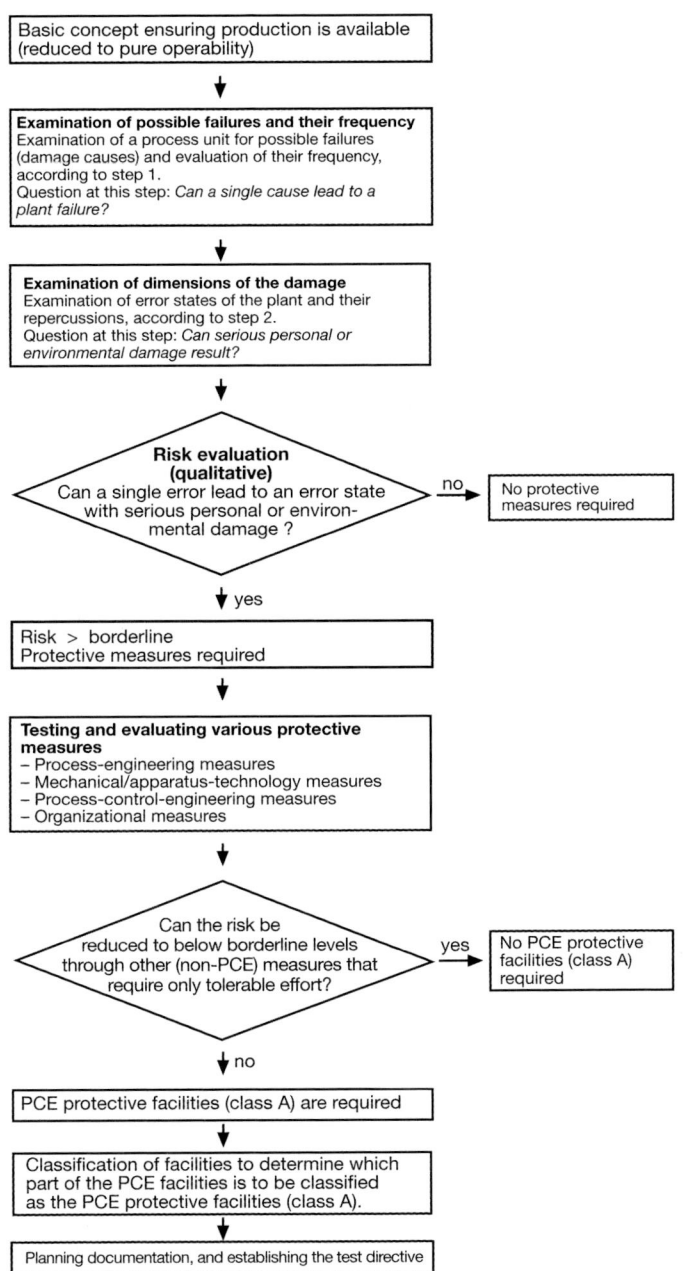

Fig. 7.6 Determining the need for protective measures

Faults Related to Product
Chemical or process engineering causes such as exothermic reactions, blockages, foams, aggregation state changes, weathering influences

Operating Errors
Step 2: Examination of Dimensions of the Damage Caused by Inadmissible Faults
Such faults could arise due to:

Failure of Apparatus, Piping, Seals, due to
- Excessive/too low pressures
- Excessive/too low temperatures
- Mechanical overload
- Material overload

Erroneously Opened Plant Discharge
This can result in severe personal or environmental damage, due to:

Release of Dangerous Substances
- Toxic substances with acute danger to humans
- Flammable substances with fire- and explosion danger and dangerous to people
- Environmentally unsafe materials with serious consequences

Mechanical Effects with Danger to People through
- Shock waves
- Scattering of rubble

Supplementing Chapter 10, we will here briefly introduce the use of process control facilities that, in the context of the above-shown concepts, serve the safety of technical process plants in the chemical industry. The objective of such special process-control-engineering safety facilities is to prevent personal injuries and significant damage to property and the environment (see also NAMUR recommendation NE 31 [7.6]). The function of a process-control-engineering protection facility (class A) is to prevent an *inadmissible* error in the plant. Their function differs significantly from that of process-control-engineering operating- and monitoring facilities.

Process-control-engineering safety facilities monitor a numerical value expressing process safety by comparing it to the allowed values; when the monitored value deviates from the allowed values, they set off a switch sequence or an alarm to alert the operating personnel to carry out the necessary previously determined organizational measures.

During interdisciplinary safety discussions, the requirements (protection objectives, function, principle of technical design, type and frequency of performance checks, other organizational measures) of process-control-engineering protective facilities are determined and documented.

Class-A Process-control-engineering protective facilities should be designed and run in a way that ensures that even if a passive fault regarded as probable occurs, the protection function will be performed anyway.

Class-A process-control protective facilities are therefore generally of redundant design (usually on the one-of-two or two-of-three principle). An additional diversity reduces possible systematic errors.

Hard-wired program control is usually used for class-A process-control protective facilities.

With complex protective facilities, the use of a process automation system or a programmable logic controller may be more economical. Certified systems may be used for single- or multichannel protective devices depending on the permit. Systems that are noncertified, but well proven in practice, may have at the most one channel of a multichannel protective device going through the process automation system or programmable logic controller.

NAMUR recommendation NE 31 [7.6] may, again, be consulted for further information on safety-related use of process control engineering.

Typical problems associated with multiproduct plants are those related to the choice of sensors (function, measurement range), discussed in Section 7.3, as well as the management and adjustment of limiting values (also especially with automated systems with sequence control/recipe packages). The details and main points related to plant safety in multiproduct plants with regard to their flexibility requirements are discussed more closely in Section 10.4.

7.6 References

[7.1] MAGIN, R., WÜCHNER, W., *Digitale Prozeßleittechnik*, Vogel, Würzburg, **1987**.

[7.2] PEINKE, W., *Entwicklung der Prozeßautomatisierung in der Chemie*, R. Oldenbourg, Munich, **1995**.

[7.3] NAMUR ("Interessengemeinschaft Prozessleittechnik der chemischen und pharmazeutischen Industrie," User Association of Process Control Technology in Chemical and Pharmaceutical Industries). Worksheet NE58, *Abwicklung von qualifizierungspflichtigen PLT-Projekten*, available from NAMUR offices, **1996**.

[7.4] DANNAPEL, B., et al., *Automatisierungstechnische Praxis 10, Qualifizierung von Leitsystemen: Ein Gemeinschaftsprojekt von GMA und NAMUR zur Validierung*, R. Oldenbourg, Munich, **1995**.

[7.5] TETZLAFF, R. F., *GMP documentation requirements for automated systems, part I–III, Pharmaceutical Technology*, **1992**.

[7.6] NAMUR recommendation NE31, *Anlagensicherung mit Mitteln der Prozeßleittechnik*, available from NAMUR offices, **1993**.

[7.7] NAMUR status report 1995, *Prozeßleittechnik für die Chemische Industrie*, a) Willems, E., *Instandhaltungsaspekt bei der PLT-Planung*, b) DRESSLER, T., TRILLING, U., *Instandhaltung in der Prozeßnahen Technik. Cost of ownership*, R. Oldenbourg, Munich, **1995**.

[7.8] STROHRMANN, G., *Einführung in die Meßtechnik im Chemiebetrieb*, R. Oldenbourg, Munich, **1987**.

[7.9] WINNACKER, KÜCHLER *Chemische Technologie 1984, 1*, offprint.

[7.10] STAAB, J., *Industrielle Gasanalyse*, R. Oldenbourg, Munich, **1994**.

[7.11] TÖPFER, H., *Automatisierungstechnische Praxis 34, Auf dem Wege vom einschleifigen Regelkreis zur universellen Leittechnik*, R. Oldenbourg, Munich, **1992**.

[7.12] NAMUR recommendation NE 28, *Empfehlung zur Ausführung von elektrischen Steckverbindungen für die analoge und digitale Signalübertragung an Labor-MSR-Einrichtungen*, available from NAMUR offices, **1993**.

[7.13] SCHNEIDER, H.-J., *Automatisierungstechnische Praxis 35, INTERKAMA 92: Sensorsysteme und die Kommunikation im Feld*, R. Oldenbourg, Munich, **1992**.

[7.14] WÜRSTLIN, D., et al. (NAMUR group 2.4), *Automatisierungstechnische Praxis 4, Anforderungen an die Prozeßleittechnik in der verfahrenstechnischen Forschung und Entwicklung*, R. Oldenbourg, Munich, **1993**.

8
Process Operation

The process operation of a continuously operated single-line plant for the production of a main product with a few side products can be very demanding. Fast startup and shutdown of the plant and running at full capacity can here make a significant contribution to the economic success of the operation. To satisfy the increasing degrees of freedom found even in the most common standard multiproduct plants (MPPs) already cause difficulties. This is even more so with the other multiproduct plant types. Whereas more economic production is the primary goal of better process operation in continuously operating single-line plants, process operation methods are only aimed at achieving basic operability in multiproduct plants. Process operation methods are therefore essential in multiproduct plants. This chapter will deal with the basic techniques of process operation in the different types of multiproduct plants.

The term *process operation* is used for the sum of all the measures taken to ensure that the production goal of the plant is achieved. Successful operation is generally not easily achieved, since the integration and multiple use of energy- and material flows, the need for consistent high productivity, high operability, high flexibility, and product quality, and conforming to standards and legal requirements often place contradictory demands on the running of the plant. Increasingly stringent requirements of the above-mentioned factors necessitate suitable process operation strategies, especially where there are no other boundary conditions, to ensure the competitive edge of the production process. If, for example, the manufacture of a product is not protected under patent law and there is no other way of ensuring an advantage over the competition, process operation becomes one of the most important competitive factors.

It should also be noted that the organization and operation of process-engineering production plants differ fundamentally in several aspects from the manufacturing processes of mechanical production already described in the literature. A batch-oriented, discontinuous process engineering operation, for example, can seldom be interrupted, interim storage between individual reaction stages is usually not possible, or available, and capacities of all kinds are generally limited. For this reason, process-engineering production will first be subjected to a structuring process in this chapter.

8.1
Model of Levels of Chemical Engineering Production Technology

Chemical engineering production technology can, for easier comprehension, be arranged into horizontal levels (Fig. 8.1, reference [8.1]). Every level summarizes functionally similar duties. Intensive information exchange within the levels is often characteristic. The lowest level is here the process itself, in its technical function. The field level above that obtains information about the process through measurement sensors. The set points are obtained by the actuators in the opposite direction. According to this classification, the measurement- and elementary control functions themselves are allocated to the field level. The next level is the process control level. Here measurement signals are processed, visualized, registered, and processed in a form relevant to the process. The set points for the controllers and actuators in the plant ultimately result from this processing. The field- and process control levels together are in the first place responsible for the elementary functions of the process control engineering. It is here where, for example, levels of containers are maintained, temperatures are controlled, but also recipes are followed and bolting is monitored. For the purposes of this chapter, continuously and discontinuously operated plants as well as all types of multi-product plants are regarded to be the same with regard to the nature of their tasks and the basic functionality required for performing these; these characteristics are already present in most of the presently available process control systems, albeit in different forms (see also Chapter 7).

The functions of process- and plant operation mainly belong to the next higher level, the operation- and production management levels. This is where the question of which product should be produced at which time in which plant section is dealt with. It is easy to see that the nature of this question depends on the type of multiproduct plant. If, for example, a continuously operated multiproduct plant is considered, the problem to solve is that of campaign planning. This includes the

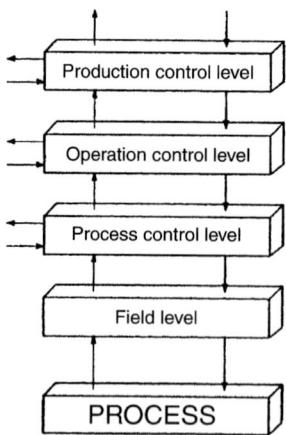

Fig. 8.1 Production levels

size (duration) of the individual campaigns, as well as the sequence of events, and, in case of a multiple-line plant, also the allocation of a campaign to a line. What needs to be dealt with is, in other words, an allocation problem, the aim of which is to keep production costs low, under consideration of changeover and cleaning times, the production of "off-spec" materials on changeover, as well as the influence of storage on capital binding. Problems of this nature have already been addressed comprehensively in the literature (e.g., reference [8.2]) and will not be discussed in further detail here.

One of the functions of management considerations in discontinuously operated multiproduct plants is to draw up optimal occupancy plans, aiming at maximizing production throughput in the relevant multiproduct plant, with the economic marginal conditions taken into account. The nature of this task depends on whether the plant is a modular multiproduct plant, a multiproduct plant with pipeline manifolds, a multiproduct plant with multipurpose apparatus, or a pipeless plant. Even where only slightly different products move through the plant along different routes, the planning task assumes a size and degree of complexity that is no longer accessible by manual calculations. For this reason, the first step in solving management- and process-operation tasks is to formally describe and thereby structure not only the plant itself, but also the recipes describing the process. It is this formalization that makes computer-aided plant allocation possible [8.3]. Formalization is furthermore essential for establishing the recipes described in Section 8.4 and for carrying out recipe-based production.

8.2
Structuring the Description of a Multiproduct Plant

Of the various standards dealing with the nomenclature of plants, processes, and recipe operations, DIN IEC 1512 [8.4] (DIN, Deutsche Industrie Norm, German industrial standards; IEC, International Electrotechnical Commission) will be considered more closely here. The explanations will at first deal only with the standard multiproduct plant as introduced in Section 3.1. In the sections following that, other multiproduct plant types will also be covered.

In Fig. 8.2, a *process cell* is first subdivided into *units*. A unit consists of one or more *equipment modules*, and these, in turn, are made up of one or more individual *control units*. The concept *process cell* therefore encompasses everything that is required to run one or more batch(es).

The structure shown can also be followed in the opposite direction, to higher levels. Although these further levels contain no aspects special to multiproduct plants, they are of interest because of their production-logistical relevancy to the site as a whole.

A frequently encountered plant type is the multiline plant. In such a plant, the charge moves from one unit to the next, whereby not all available units need always be used and several lines may operate in parallel. The sequence in which the units are used during the production of a batch is called the *batch route*. In

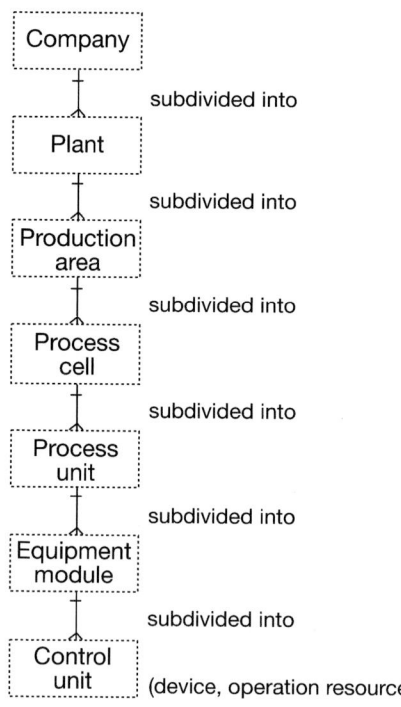

Fig. 8.2 Plant-structuring concepts

multiproduct plants with pipeline manifolds, several route variations along the batch path and also partly between units and the equipment modules assigned to them are made possible by the flexible application of pipelines. In a multiproduct plant with multipurpose apparatus, in contrast, transport of a charge from unit to unit is often not even necessary (this is often deliberately avoided because of troublesome material properties, e.g., extremely high viscosity or solid ingredients), as the apparatus itself is very versatile and therefore dispenses with the need for transport. In a pipeless plant, on the other hand, the term *unit* assumes a completely different meaning, as a unit – the mobile container – moves from equipment module to equipment module and thereby describes the batch route. The unit, therefore, with the exception of the stirrers, is considered and allocated without its equipment modules. The equipment modules are therefore regarded as only temporarily available and temporarily occupiable resources.

A *unit* (*plant unit*) is made up of *equipment modules* and individual *control units*. They may be used only part of the time. A unit is typically responsible for the larger process activities, the so-called process stages, such as the precipitation of a product, the evaporation of a solvent, or a reaction, and is as independent of other units as possible. Larger process entities such as a stirred vessel or a reactor generally correspond to a unit. In a multiproduct plant with multipurpose apparatus, the unit is supplied with unusually many and versatile equipment modules. In the pipeless plant, the unit is generally not connected to its equipment modules and individual control units. There are exceptions to this rule; stirrers are, for example, sometimes permanently installed in mobile vessels.

An *equipment module* is basically made up of *individual control units*. An equipment module often comprises an independent group of apparatus within a unit, and is thereby part of that unit. Filters, evaporators, and condensers may be considered examples of equipment modules. Equipment modules may be used exclusively or in parallel. They are typically responsible for tasks lower on the hierarchical scale, such as the dosing of precipitation agents or the condensation of vapor. Barring a few exceptions, there is no difference between the equipment modules of standard multiproduct plants, modular multiproduct plants, multiproduct plants with pipeline manifolds, and multiproduct plants with multipurpose apparatus. The equipment modules of a pipeless plant have fixed locations and are only temporarily assigned to a unit.

From a technical control point of view, a *control unit* functions in a self-contained manner and is therefore considered to be independent. Measurement- and control units at any level are typically regarded as control units. A regulator consisting of flow transmitter, regulating valve, and actual flow regulator is, for example, regarded to be a control unit. Another example is the automatically operated switch valves of a dosing unit, which direct the flow in one or more predetermined direction(s). The specific design of the sensors and actuators may, however, differ substantially from case to case, especially in pipeless plants (see also Section 7.3).

8.2.1
Structuring the Standard Multiproduct Plant and the Modular Multiproduct Plant – An Example

Fig. 8.3 shows an example already known from previous chapters, that of a standard multiproduct plant. The section of the plant shown probably does not represent the functionality required to run the whole process, only a part thereof. In the top part of the figure, three batch vessels are shown. These can be used for the preparation of starting materials and the subsequent dosing of these into the reaction vessel below them. Because they may contain a substantial part of the charge, and large "process stages" may therefore take place in them, these batch vessels are structured as units. In this figure, one equipment module, that of dosing, has been allocated to each unit.

If these three batch vessels are equipped with the same instrumentation and they are identically equipped with individual control units for their dosing function, they may be allocated flexibly. This flexibility may, however, be limited by special stirrers, variable dosing unit designs, or different vessel volumes. Such special details also need to be taken into account during planning.

Fig. 8.4 shows the concept of a modular multiproduct plant. Two batch vessels are shown in the top half. In contrast to those in Fig. 8.3, these are flexibly connected by tubing, in other words, the connections between the units are specific

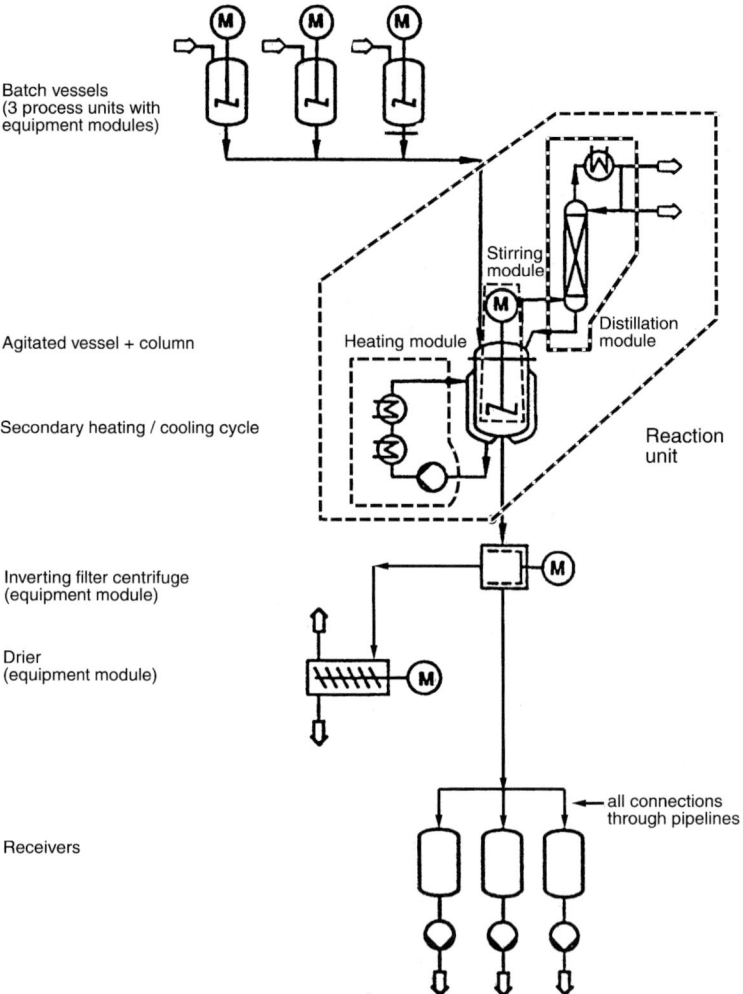

Batch vessels
(3 process units with
equipment modules)

Agitated vessel + column

Secondary heating / cooling cycle

Inverting filter centrifuge
(equipment module)

Drier
(equipment module)

Receivers

Stirring
module

Heating module

Distillation
module

Reaction
unit

all connections
through pipelines

Fig. 8.3 Part of a standard multiproduct plant with some equipment modules

to the current batch. This extends the planning question to equipment modules
beyond the limits of the unit – possibly even beyond the limits of the whole plant.

If the batch vessels are only used for dosage, the choice of their allocation is
open, since they no longer need to be regarded in combination with their attached equipment modules, as is the case with standard multiproduct plants. In
contrast, here the dosing equipment modules assume the character of limited resources, for which the different units compete.

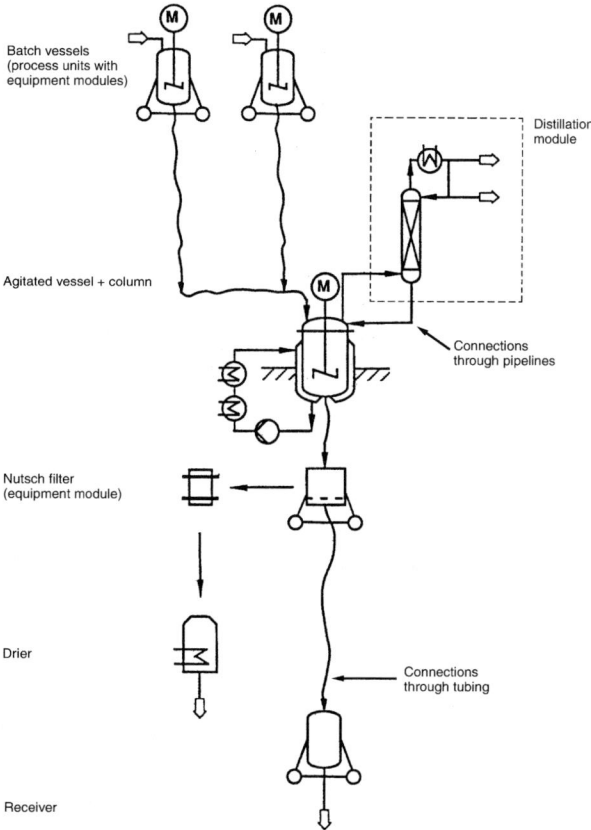

Fig. 8.4 Part of a modular multiproduct plant with some temporarily allocated equipment modules

8.2.2
Structuring the Multiproduct Plant with Pipeline Manifolds

In Fig. 8.5, a number of multiproduct plant units with pipeline manifolds are shown. The section of the total plant shown in this figure is larger than that used in the discussion in the previous sections. That is why three process units, vessels with mounted columns, are shown in this figure. In this case, the process units also compete for equipment modules such as filter centrifuges or driers. In modular multiproduct plants, and also in pipeless plants, described next, some of these equipment modules are mobile. However, because of the generally fixed floor plan, they cannot be regarded as freely available resources. Above the vessels, the dosing equipment modules are available in various forms. If the content of one of the batch vessels is dosed or filled into another agitated vessel, this will proceed over the manifold, also known as a "dosing spider." The manifold, depending on

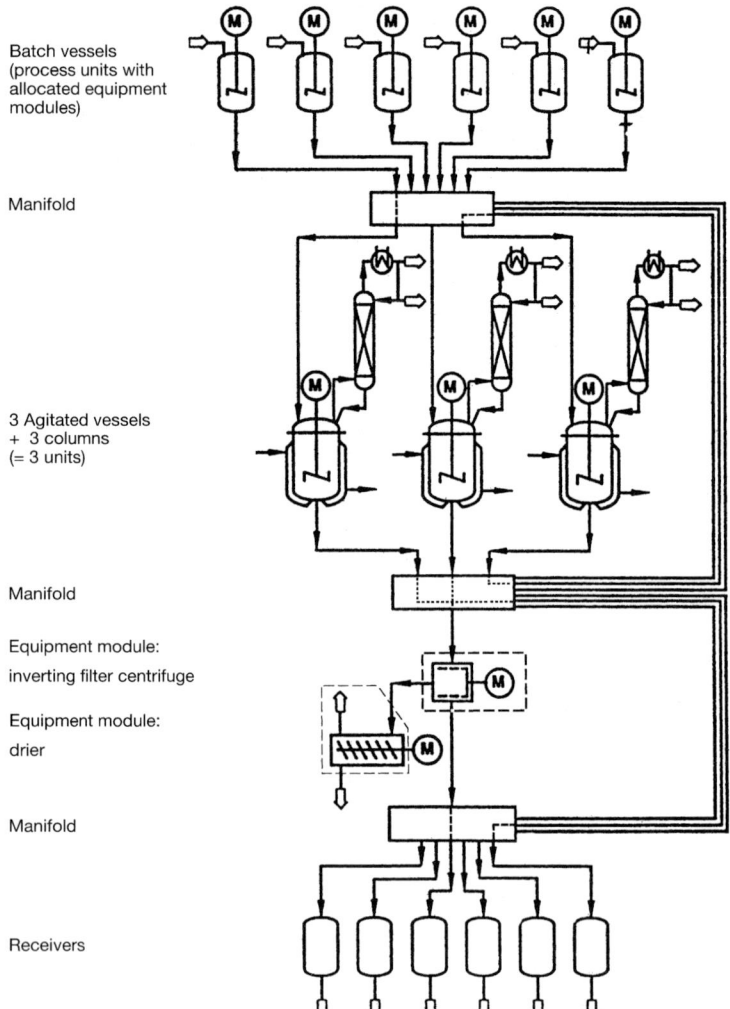

Batch vessels
(process units with
allocated equipment
modules)

Manifold

3 Agitated vessels
+ 3 columns
(= 3 units)

Manifold

Equipment module:
inverting filter centrifuge

Equipment module:
drier

Manifold

Receivers

Fig. 8.5 Multiproduct plant with pipeline manifolds and selected equipment modules

its specific design, can therefore function as an equipment module that is either exclusively allocated (with one outlet) or in-parallel (with more outlets).

Below the agitated vessel another manifold is located. With this, flow is possible in the opposite direction, back from the lower to the upper manifold, thereby allowing transfer of the content of one vessel to another. If the different vessels are equipped with different equipment modules and/or individual control units, multistep syntheses with different synthetic step requirements can be run in the different process units.

The capacity of the manifolds should be calculated carefully to ensure that enough outlets are available at all times, to avoid the operation of the plant at its extreme limits with regard to use and throughput. Even if every loading or dosing operation results in a manifold outlet being occupied, time-related boundary conditions (the charge has to be let out *right now*) or emergency shutdown procedures may under no circumstances be hindered. This should, of course, be taken into account for recipe operations.

8.2.3
Structuring the Pipeless Plant

In Fig. 8.6, a process unit of a pipeless plant is represented. A vessel on a mobile frame with an equipment module responsible for stirring is shown in the center. The stirrer is the only equipment module that is permanently allocated to the vessel and therefore poses no scheduling problem. Temperature control at the station is achieved by connecting pieces automatically fitted to the illustrated double casing or through welded-on piping. In contrast to the equipment described so far, the rest of the functions are provided by dedicated stations. Some of the stations are units with dedicated equipment modules and others are purely equipment modules that are only temporarily allocated to the vessel. In terms of allocation planning, every station is treated separately. How the allocation problem is dealt with at each individual situation depends, among others, on the specific technical process being carried out. If, for example, the reaction system under consideration has lower but no upper time limits, the stations can be regarded as resources in a queue system. The mobile container then docks at whatever station is next in the queue. If, however, the reactions require exact time limits to be kept, queuing is impossible. It should then be ensured, for example, by careful timing of occu-

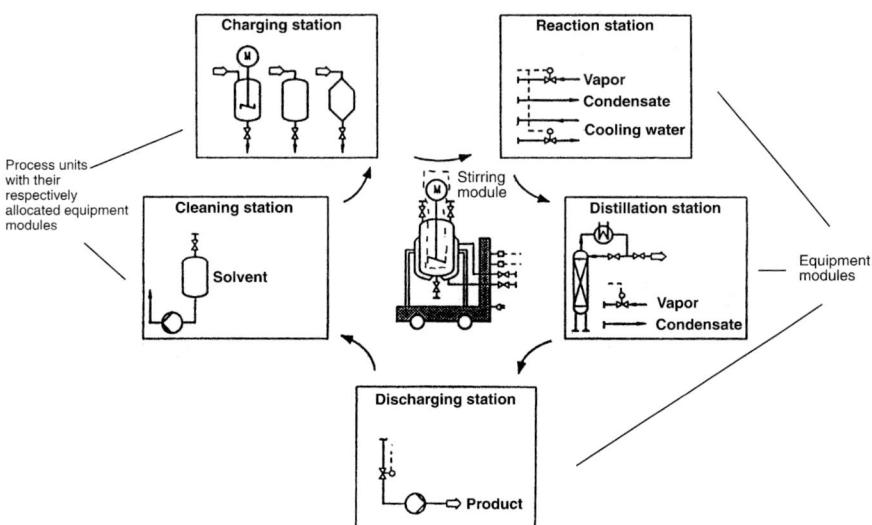

Fig. 8.6 Section of a pipeless plant with some plant units and equipment modules

pation, that the required stations are always available when needed. The individual control units permanently allocated to the respective equipment modules in pipeless plants do not differ from those of other multiproduct plants.

The structural flexibility (as defined in Chapter 1) of especially the plant types discussed just now is far superior to that of standard multiproduct plants. On the one hand, this is positive, as it enhances the applicability of the plant: a broader product spectrum becomes accessible and the plant can generally operate to fuller capacity or with greater throughput. But the down side of this is that the allocation and scheduling problem becomes much more complicated and of a totally different nature. It is therefore important already in the conceptual stage to use suitable tools (Chapter 9) so that potential bottlenecks are recognized and removed as early as possible.

8.3
Structuring the Description of Batch Processes

A *process* is a sequence of chemical, physical, or biological events from which, in the end, a new or altered product is obtained. Because this statement is very general, it can be formulated completely independently of the plant type in which the process will run at a later stage. The amount of product resulting from a single batch process is called a *batch*. Batch processes are discontinuous. They have the characteristics of both discrete processes (e.g., discrete manufacturing processes) and continuous processes. In Fig. 8.7, the hierarchical breakdown of the total process into *process stages* is shown. The latter can, in turn, be subdivided into *process operations*, which consist of a specific number of elementary *process steps*.

The various *process stages* making up a process can proceed independently in series, in parallel, or in a mixed form, and generally relate to a chemical or physical transformation. Typical process stages are, for example, a polymerization reaction, the evaporation of a solvent, a concentration measurement, or the granulation of

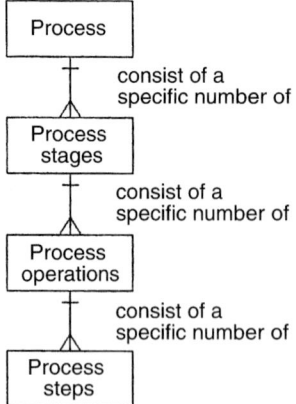

Fig. 8.7 Process structure

a powder. Process stages are separate from each other. Materials with defined properties make up the connections between the process stages.

A process stage is made up of one or more closely related *process operations*. These describe comprehensive process activities, but not the complete transformations found in process stages. A typical process operation of a process stage could, for example, be the introductory evacuation of a vessel, the removal of a sample for a concentration measurement, or the dispensing of a specific amount of distilled water.

The *process step* is the smallest unit of the hierarchy shown in Fig. 8.7. Process steps typically describe small procedures, for example, the adding of a small amount of catalyst or heating to a specific temperature.

8.4
Recipe-Based Operation

A recipe is defined as the complete and unambiguous set of instructions containing the minimum amount of information necessary for the production of a specific product. A recipe therefore contains neither the allocation planning, nor the selection of the specific equipment; it deals exclusively with the product. Important for the practical application of a recipe is the fact that the different parts of a firm need information with different levels of detail about the production process. For this reason, there are different types of recipes. A hierarchical description is given in Fig. 8.8.

The *process recipe* is completely neutral with regard to its implementation. The recipes at the lower levels are derived from the process recipe. This recipe is still completely independent of the plant in which it is to be run; more specifically, for the process recipe, the type of multiproduct plant in which it is implemented is irrelevant. The quantities given may be defined or relative values. Equipment requirements are only indirectly defined through the process conditions, for example, pressures or temperatures. The process recipe is therefore not only an instrument for planning related to the firm as a whole, but it is also a communication agent between the process-defining and the process-implementing disciplines in chemical industries. The *works recipe* is a more detailed, location-specific recipe, and, in the context of this chapter, does not differ from the process recipe.

One or more *basic recipe(s)* are derived from the works recipe, and contain, for the first time, plant-specific information. The basic recipe is already sufficiently adapted to the properties of the equipment to ensure proper production. It therefore contains everything required for detailed production planning. The basic recipe is, however, still relatively neutral with regard to the required process units. Although the type of multiproduct plant and an approximate plant structure regarding the required capacities and equipment modules are already implied, the individual units have not been allocated to specific batches yet. A basic recipe characteristically contains the names, quantities, and properties of the feedstocks and products, and describes the production aims in a textual form.

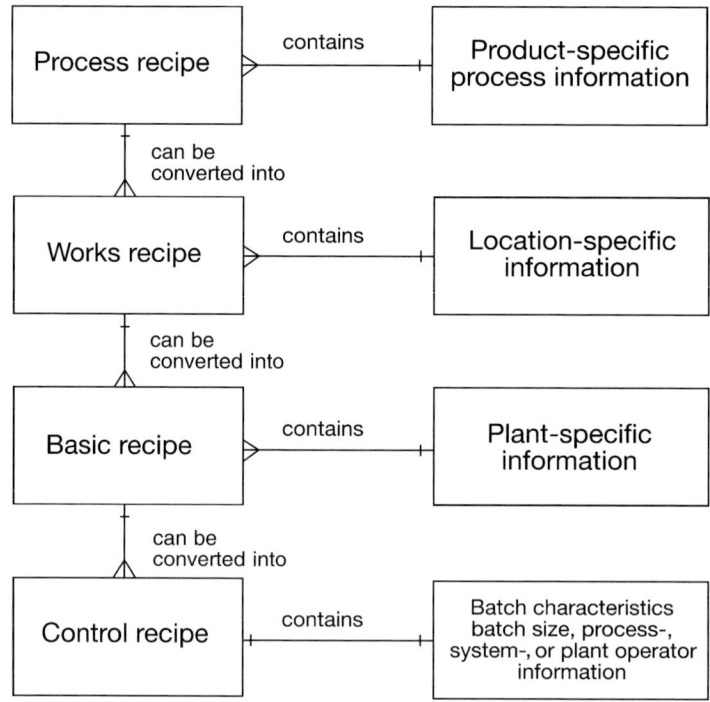

Fig. 8.8 Recipe types

The *control recipe* is an expanded version of the basic recipe, and contains allocation and operating instructions for a specific batch. In this recipe, the information from the structural description of the plant and the basic recipe come together. For example, the units to be used are mentioned and assigned, and exact quantities are given. The control recipe should also give information on the measures to be taken in case of irregularities and deviations from the defined operating conditions.

8.5
Orientation of Recipe-Based Operation by the Model of Levels of Process-Engineering Production

Fig. 8.9 shows the standard structure for recipe operation recommended by NA-MUR ("Interessengemeinschaft Prozessleittechnik der chemischen und pharma-zeutischen Industrie," User Association of Process Control Technology in Chemical and Pharmaceutical Industries) [8.5]. It is based on the model of levels of production (see right-hand edge); it shows how the different disciplines collaborate during process-engineering production and is also a useful summary of the information given in this book section. The functions at the operation- and process control levels are of major importance for recipe operation.

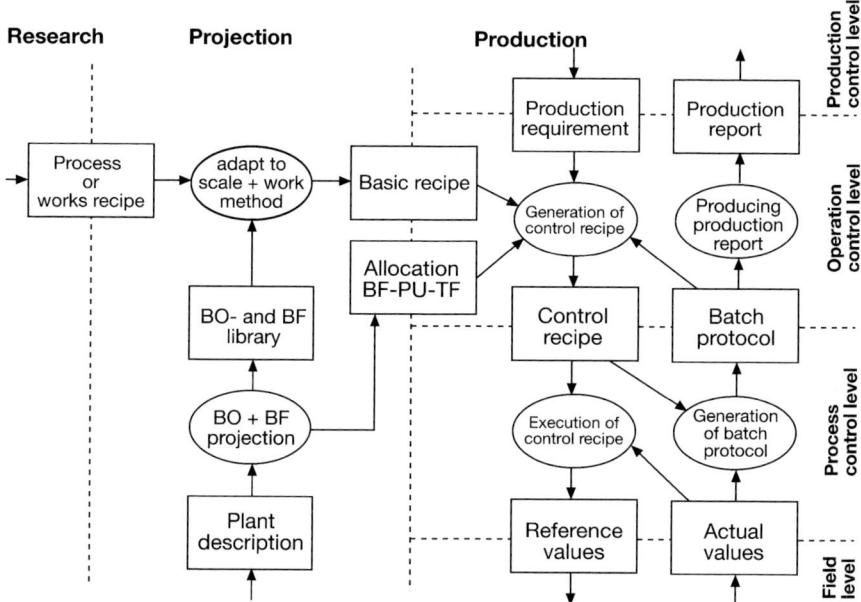

Fig. 8.9 Standard structure of recipe-based operation. BO, basic operation; BF, basic function; PU, process unit; TF, technical function

The process- or works recipe (Fig. 8.9, left-hand side) is generated by the research and/or development department(s) of a firm. This recipe is not yet very formalized, but this does not substantially hinder communication in the earlier phases of process development.

During a projection stage, the formalized description of the plant (Section 8.2) is generated with a library of basic operations and basic functions that can be used in that plant. Further on, this projection stage also serves to assign the basic functions to the different units and their technical functions. If this description of the technical functions is then stored in the process control system, the information on *how* the control functions from the control recipe are to be applied concretely in that unit is then found in a form that can be processed automatically.

With the library of basic functions taken into account, the basic recipe can now be generated. The basic recipe, as well as the allocation of basic functions, units, and technical functions derived from the plant description, is now the actual interface with the production process. From the description of the technical functions already stored and the basic recipe it is possible, directed by the production requirements set by the production management level, to automatically generate the necessary unit allocations and, following this, to automatically generate the corresponding control recipes.

It is easy to recognize the most important advantage of this method: if the plant has been completely structured once and the process has been formalized in the form of a recipe, production tasks can be automatically worked out for a complex plant structure; this eliminates the need for manually choosing and allocating the plant units with their technical functions for every separate case.

When the control recipe is carried out (Fig. 8.9), controller set points that should be conformed to are generated at the field level. From the feedback of the actual values and the control recipe itself, the batch protocol can be generated. The batch protocol in its turn is condensed into the production report, which is responsible for continuous information feedback at the production management level and also beyond.

8.6
Styrene Polymerization in a Multiproduct Plant

Styrene can be polymerized in several different ways, to give products with widely different properties. Here the production of expandable polystyrene in the various multiproduct plants will be examined from the point of view of efficient process operation. The focus will be the application of the structuring principles for processes and plants introduced in the previous sections and not so much to provide a realistic scenario. Also note that, for space reasons, the specialized terms relating to polymerization technology are not covered here. They are, however, easily found in the literature.

For the styrene polymerization considered here, the monomer is dispersed as small droplets in water. The low solubility of the monomer in water makes it very suitable for this type of polymerization. The main factors influencing this polymerization are temperature, stirrer type, stirring rate (thus, droplet-size distribution), type of initiator, type of dispersant, and reaction time.

At the temperature of the reaction tank, the fine monomer droplets gradually react. As soon as approximately 30% to 70% of the monomer has been converted, the monomer droplets become sticky because of the polymer molecules dissolved in them. This causes droplets coming into contact to agglomerate, a process that can lead to complete coagulation of the entire tank content, unless effective countermeasures have been taken.

To prevent such agglomeration, solutions of a range of salts and so-called protection colloids are added to the reaction mixture. These prevent this coagulation and at the same time stop the monomer from emulsifying.

If the polymerization has been conducted properly, no monomer residues should remain, as these would have a negative effect on the properties of the final product. The absence of monomer residues is ensured by the use of various reaction initiators (e.g., peroxides) with different decomposition characteristics and by a stepwise increase in reaction temperature.

If the polystyrene should be expanded into a foam (e.g., to Styropor), pentane is added as the foaming agent to the reaction mixture. Various precautions are necessary for the handling of such gaseous foaming agents, and these are expressed in the requirements placed on the apparatus that is used for the process.

8.6.1
Structuring an Existing Plant

Fig. 8.10 is a diagrammatic overview of the standard multiproduct plant in which our styrene polymerization is to be performed. Shown are various containers labeled B..., reaction vessels labeled R..., and a centrifuge labeled S.... In Fig. 8.11, the *process cell* is shown in more detail. Here the two *plant units*, vessels R814 and R815, can be distinguished. Fig. 8.12, in its turn, shows the detailed periphery of vessel R811. This vessel is shown with various *equipment modules* for dosing as well as equipment modules for stirring and discharging. The *basic control unit* of valve HV8113 seen in Fig. 8.12 is, in turn, shown in Fig. 8.13 with the state of the different sets of equipment given in greater detail. This structuring process should be carried out to the same extent for each set of equipment in the process cell.

Already in the relatively small and clearly laid-out plant shown here, it becomes obvious that a certain amount of complexity is involved in the structuring process. This makes its formal execution even more important.

The final step in the structuring of the plant is the projection of the library of basic operations or basic functions. This is where, for example, the information is found on how the equipment module "dosing" is to be implemented in each specific case. This, of course, generally differs for the different types of multiproduct plants and even for the specific plants or units. For example, for the equipment module "dosing" in a multiproduct plant with pipeline manifolds, it should be ensured that the corresponding outlets are available in the dosing spider. In a modular multiproduct plant, the equipment module "dosing" should contain information on whether additional equipment for the unit, such as appropriate piping, has been supplied in time.

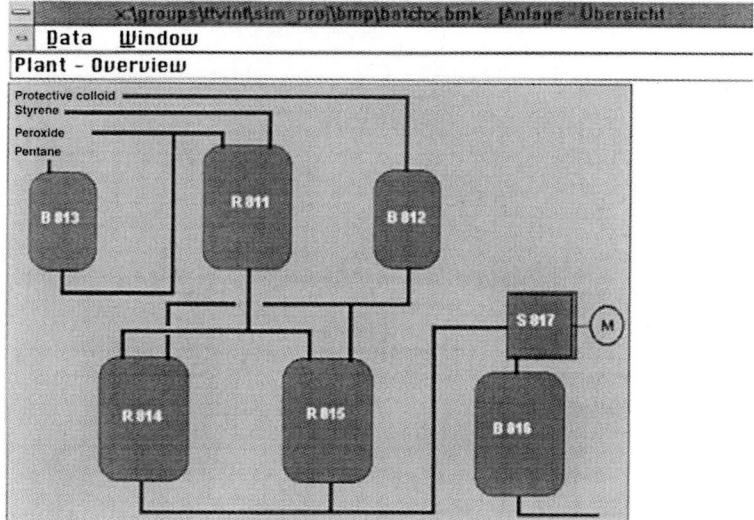

Fig. 8.10 Polymerization example, overview

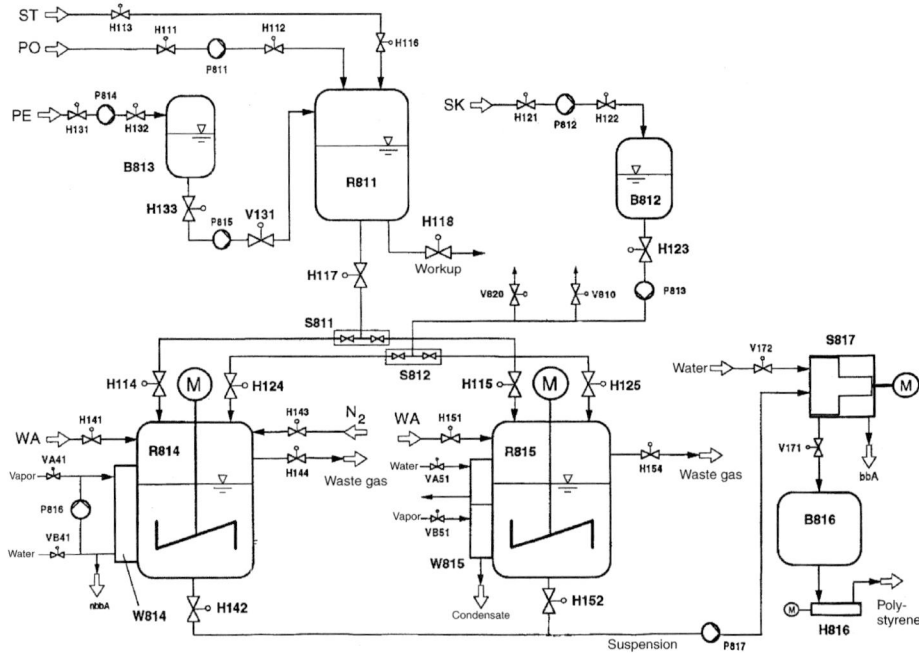

Fig. 8.11 Polymerization example, reactors

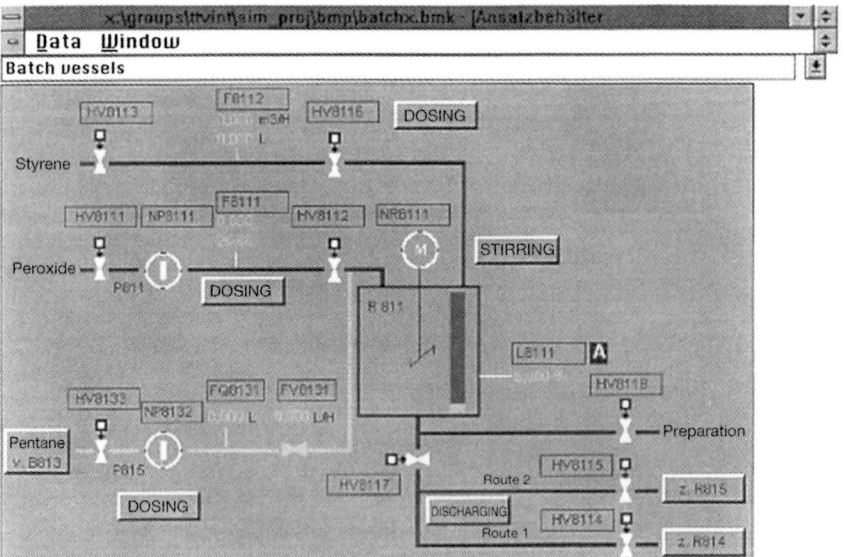

Fig. 8.12 Polymerization example, batch vessels

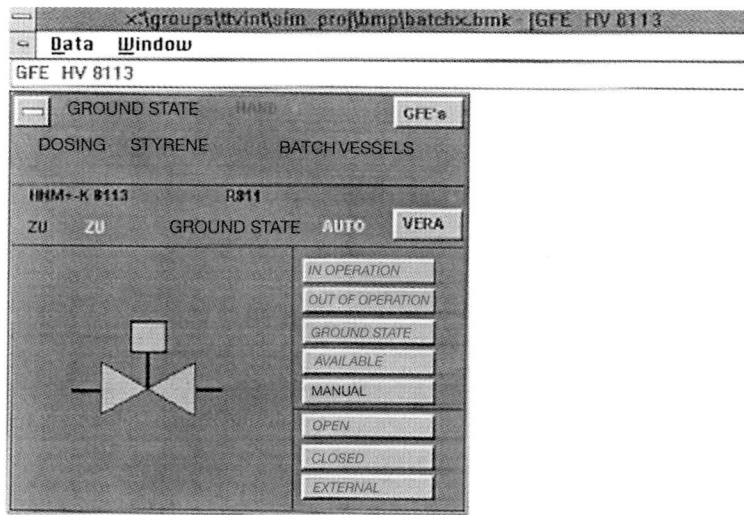

Fig. 8.13 Polymerization example, basic control unit

The library created with this has a double function. Firstly, it is utilized during recipe operation for the conversion of the basic recipe into the control recipe. Secondly, it is applied during allocation planning for the plant, as it can be used for determining whether the plant can, at all, meet the requirements set by a specific batch. In this way, the library of our example plant leads us to conclude that polymerization of styrene to expandable beads is possible in it, as the batch vessels are equipped to deal with gaseous foaming agents.

8.6.2
Structuring the Process

In Fig. 8.14, a section of the structuring of the process into process stages is shown in the form a of a phase model. In the subsequent Fig. 8.15, the corresponding basic recipe is shown as interlocking diagrams. Here parallel processes are synchronized according to the horizontal double lines. After the reagents have been loaded and discharged, the actual polymerization takes place. The protection colloid is added and the reaction mixture is heated, then cooled. The heating phase minimizes the residual monomer content. Finally, a sample is removed, analyzed, and, depending on the result, the polymer is then separated from the mixture by centrifugation.

Each process stage can be further subdivided into process operations. The introductory step of charging is used to exemplify this process, as shown in Fig. 8.16. For this, some operations need to be carried out simultaneously, for example, dosing and stirring, but sequential operations are also shown, such as "dosing" and "dosing 1." The structuring steps shown here are to be carried out for the entire process.

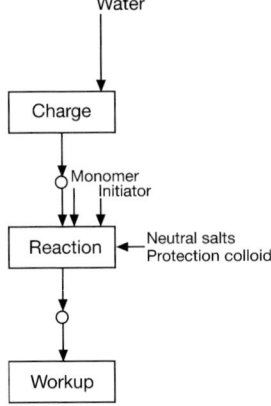

Water

Charge

Monomer
Initiator

Reaction ← Neutral salts
Protection colloid

Workup

Fig. 8.14 Process example, phase model with process stages

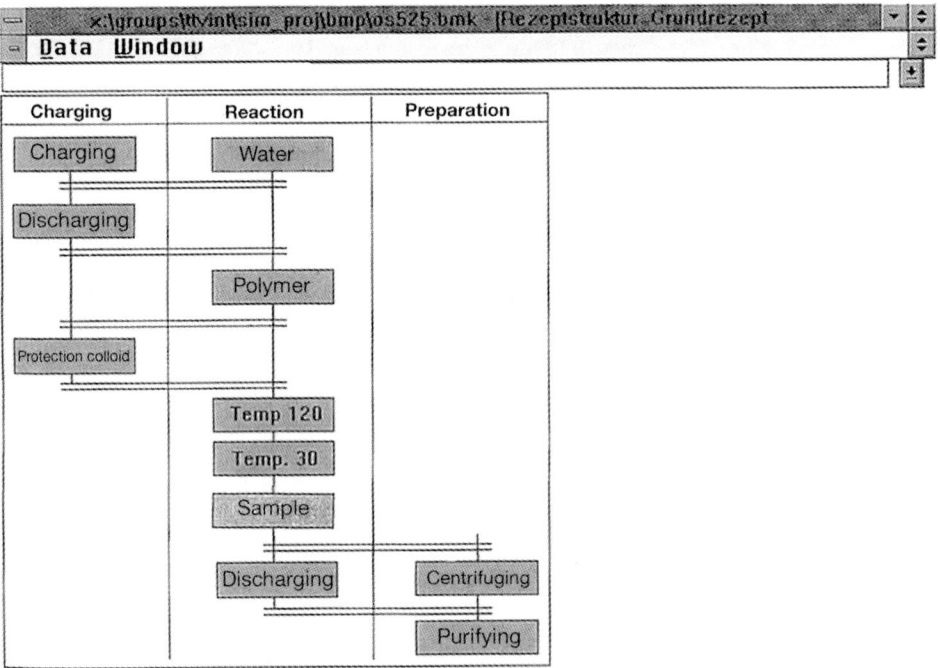

Fig. 8.15 Process example, basic recipe

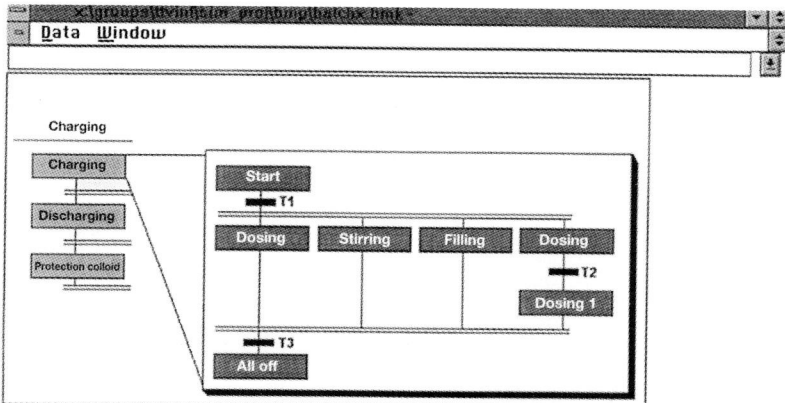

Fig. 8.16 Plant example, structuring the basic operation of "charging"

It is easy to recognize that in this abstract form of process structuring, one does not need to take the actually available units into account. It is not, for example, necessary to indicate how "dosing 1" should actually be implemented and carried out.

8.6.3
Recipe-Based Operation

It is during the run (recipe operation) that the information from the plant- and process structuring is combined. For this, the process recipe is used as starting point. A possible process recipe is shown in the following table:

Recipe heading	Designation	Polystyrene, bead-type
	Description	Production of...
	Version	1.2
	Standard scale	1328 kg
Feedstocks	Styrene	x kg
	Water	w kg
	Pentane	z kg
Products	Polystyrene	1328 kg

Apart from the exact designation and description of the styrene polymerization process, the quantities of feedstocks and products for a standard procedure are also given. The basic recipe developed from this process recipe already contains some more information necessary for planning, for example, the time required for a standard procedure. But the basic recipe is also still very neutral regarding the concrete production plant to be used. In it, only a few basic boundary conditions (e. g., batch- or continuous operation) are given.

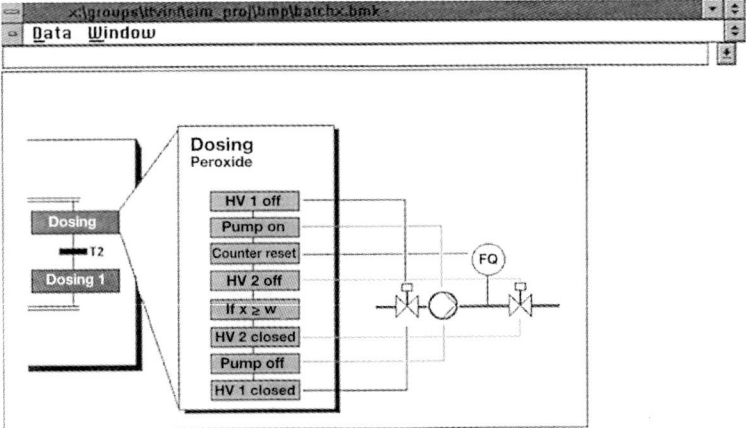

Fig. 8.17 Plant example, control recipe

The control recipe that is actually used is generated from the information from the basic recipe and the already projected library of basic operations from the example plant. Here the process-neutral library of technical functions is combined with the plant-neutral basic functions of the basic recipe. The control recipe is therefore highly specific not only regarding the multiproduct plant type, but also regarding the batch to be produced and the units to be utilized. This process is illustrated in Fig. 8.17 with the basic operation of "peroxide dosing" as example. This basic operation contains the order in which, for example, valves are to be opened, pumps are to be switched on, and counters are to be set. The control-recipe sequence shown is automatically activated. This ensures that, during production, dosing in different reaction vessels is carried out properly, in accordance with the specific equipment available.

8.7
References

[8.1] MAGIN, R., WÜCHNER, W., *Digitale Prozeßleittechnik*, Vogel, Würzburg, **1987**.

[8.2] SAHINIDIS, N.V., GROSSMANN, I.E. *Comput. Chem. Eng.* **1991**, *15(2)*, 85–103.

[8.3] KERSTING, F.J., *Funktionen der Betriebsleitebene bei chargenorientierter Produktion. Automatisierungstechnische Praxis* **1997**, *39(2)*, 13–25; *39(3)*, 56–59.

[8.4] Deutsches Institut für Normung (German Institute for Standardization), IEC 1512 (IEC, International Electrotechnical Commission), *Chargenorien-*

tierte Fahrweise, Modelle und Terminologie (developed from DIN IEC 65A (SEC) 160).

[8.5] NAMUR ("Interessengemeinschaft Prozessleittechnik der chemischen und pharmazeutischen Industrie," the User Association of Process Control Technology in Chemical and Pharmaceutical Industries). Worksheet NE33: "Requirements to be met by systems for recipe-based operations." English translation, 19 May **1992**.

9
Material-Flow Analysis by Dynamic Simulation

9.1
Meaning of Material Flow in Multiproduct Plants

To write much about the meaning of material flow in a multiproduct plant is like carrying coals to Newcastle. Material flow in a multiproduct plant plays a decisive role, because a plant can run at full capacity only if the different components are harmonized. It therefore means that material flow in a multiproduct plant should not only be examined during daily operation (production planning, process operation), but also in advance, during the planning and projection stages of a plant. Deciding whether two larger or five smaller pieces of apparatus should be used for a specific process step is certainly a technical question, but the expected material flow is also a crucial determinant.

Material flow plays an important role not only during the scaling of individual units, but also for the entire layout of a plant. Apart from the technical equipment, there are other important components of a multiproduct plant, such as buffer- and storage containers, whose dimensions need to be determined, and, of course, transportation routes need to be planned carefully. Two examples from the area of pipeless plants will be given to illustrate this point.

In the first case (Fig. 9.1), all processing units are constructed along a circular route with transport vehicles without operators. The material is transported in reaction containers from station to station. Bypass bays (P) facilitate passing maneuvers.

In the second case (Fig. 9.2), transport and processing proceed on two separate levels. On the upper level, transport containers can move in any direction towards the individual docking stations, where they are discharged and reloaded for processing in the units on the lower level.

The two layouts differ substantially. The first case is characterized by fully automated production with little buffering possibilities for the materials. The batch size is limited by the fixed structure of the transportation system.

In the second case, more buffer possibilities are available, as the number of the transport containers can simply be increased and is not correlated to the number of parking spaces. The system is more flexible and is more easily influenced from the outside. The production and procedure planning need not be fixed in as

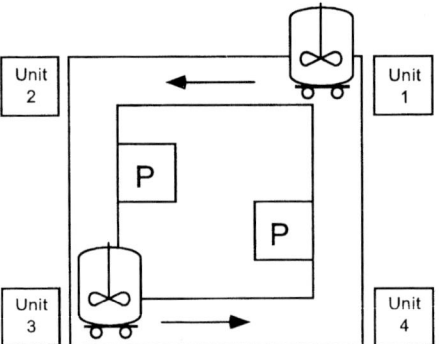

Fig. 9.1 Plant layout with circular traffic on one level

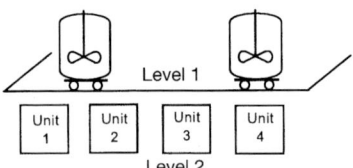

Fig. 9.2 Plant layout with separate levels

much detail as in the first example. The price is that the production in the second example is more labor-intensive.

An approach for solving such layout and dimensioning problems, increasingly encountered in the chemical industry mainly for pipeless plants, is to use a dynamic simulation model to represent the individual plant sections such as reactors, buffer vessels, and transportation systems, as well as the nonstationary material flow between the individual units and the inlet and outlet stores.

By modification of the representative models during the projection stage, the new plant as well as the individual units (apparatus and storage vessels) can be scaled, and the layout of the entire plant can be worked out. Because of the complex production procedures found in the chemical industry, especially in multiproduct plants, a dynamic material-flow simulation is the most suitable tool for material-flow analysis.

9.2
Simulation Technology

9.2.1
Thermodynamic Simulation in Process Technology

Procedure-independent process technology variables such as the process parameters pressure and temperature, the time spent in the reactor, or the suitable apparatus type are determined by laboratory and pilot-plant experiments or by thermodynamic or hydrodynamic calculations. Several methods and tools are available for solving such process-engineering problems. Especially during the last two decades, very valuable thermodynamic stationary simulators that can be applied for process development and design of apparatus have been developed. These simulators are, however, not suitable for the purposes presented here; this is also the case for the simulation approaches that are occasionally used for production planning or for drawing up planning tables.

Thermodynamic Simulators
Purpose: Design of apparatus and processes
Data entered: Material and apparatus characteristics and process parameters
Result: Stationary operation locations
Programs: AspenPlus, Hysim, ChemCad, etc.

There are no classical tools available for dynamic material-flow simulation in process technology. This is because the combination of process technology and in-plant logistics has only become important over the last few years due to increasing economic pressure. Furthermore, there is currently a move away from R&D activities of the large continuously operating plants, which can be described very well with thermodynamic simulators, to more flexible batch- and multiproduct plants. In-plant logistics play a crucial role in the latter plant types, but there are no suitable and efficient tools available for their description and analysis [9.1].

9.2.2
Dynamic Simulation in Manufacturing Technology

In manufacturing technology, numerous simulation tools have been developed for analyzing and representing the material flow, starting with the entry of the product, through the production process, up to where the finished product reaches the customer. This has come about because of the high costs associated in manufacturing processes with logistical operations such as storage and transport, where logistics and production are closely related. This connection already needs to be taken into account during the planning stage of a project. In contrast to the simulators known in process technology, the simulation tools in manufacturing technology do not examine the processes taking place within individual pieces of equipment; they are only concerned with properties related to the residence time of discrete material streams relative to several manufacturing units.

Material-Flow Simulators
Purpose: Analysis of dynamic material flow
Data entered: Residence times, flow quantities, logistical connections
Result: Equipment and machinery running to full capacity, retention times, production-planning quantities, resource consumption
Programs: SIMPLE++, Automod, Witness, Slam, Factor, PERFACT, etc.

A feature of these tools is that one can follow all the movements of the plant units and personnel on screen. This animation enables detailed on-screen analysis and observation of, for example, how material bottlenecks and shortages can

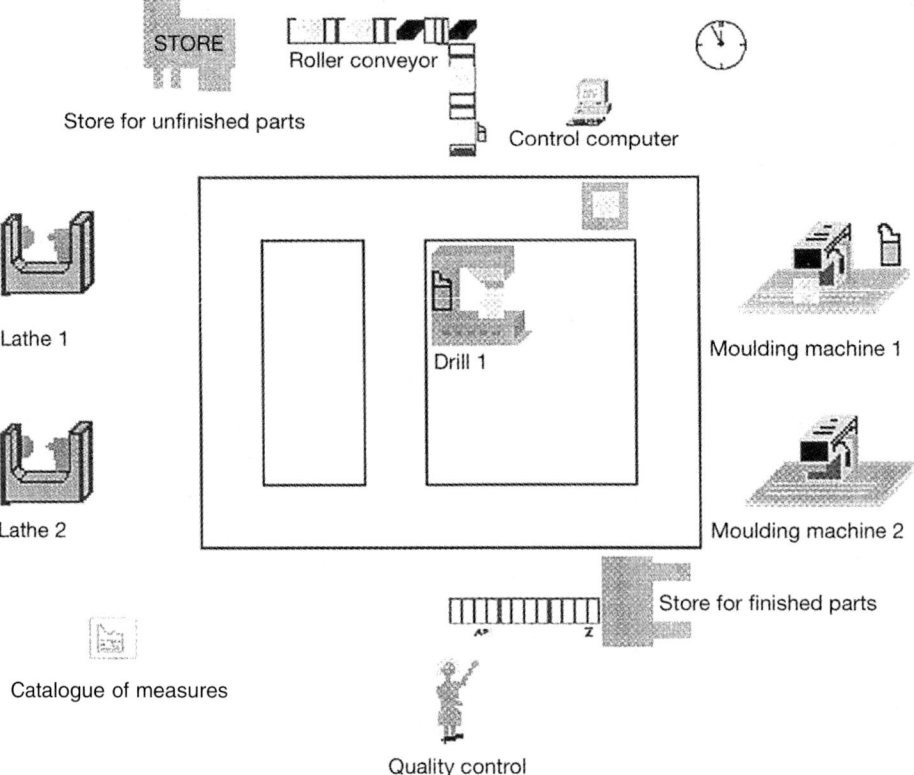

STORE

Roller conveyor

Store for unfinished parts

Control computer

Lathe 1

Drill 1

Moulding machine 1

Lathe 2

Moulding machine 2

Store for finished parts

Catalogue of measures

Quality control

Fig. 9.3 Animation surface of a workshop model [9.2]

come about on an assembly line. An example of this is shown in Fig. 9.3. Unfinished parts are transported with the aid of a roller conveyer from the store to an assembly line, where a number of manufacturing stations are found. Questions that come up in this regard, such as: "What would the effect of a second drill machine and/or another quality control point be on the production capacity?" can be answered with the help of simulation.

The VDI ("Verein Deutscher Ingenieure", German Engineers' Association) guideline "Simulation of logistical-, material-flow-, and production systems" (VDI 3633) defines simulation as follows:

"Simulation is the reproduction of a dynamic process occurring in a system; it is carried out with the aid of a model that can be experimentally tested, to result in findings that can be transferred to actual situations."

To be able to reproduce reality as closely as possible, in the sense of this guideline, it is necessary, especially in the case of process engineering, to be able to model the blurring that occurs during the process. In contrast to a manufacturing process, the residence time of a substance in a reactor is only to be indicated by a residence time spread. This is why the ability of modern simulators to work with statistical values is especially important for the chemical industry. It allows the description of fluctuating process times, breakdowns, variable batches, and so forth. It is, however, important to calculate the simulation period in a way that ensures that the given fluctuations are sufficiently backed up by statistics.

9.2.3
Difference between Process Technology and Manufacturing Technology Regarding Their Simulation

The commercially available simulation programs have been developed for use with discrete processes and the accompanying unambiguous objects. In the chemical industry, the decisive variables are, however, not single units, but continuous and discontinuous material streams. In manufacturing processes, material, time, and objects can be unambiguously assigned. This principle is valid even for transportation procedures. In the chemical industry, in contrast, several units (in Fig. 9.4 these units are the batch vessels, the piping, the pumps, and the final vessel) may be occupied by the same material for longer periods.

There are three possible ways of modeling "process engineering" material-flow characteristics while maintaining the high functionality of manufacturing simulators (statistical functions, calculations, animation):

- Adapting an existing simulator accordingly (e.g., SIMPLE ++)
- Developing a new simulator (e.g., profiSEE, BATCHES)
- Expanding the functions of existing thermodynamic simulators (e.g., planned at AspenTech).

9.2.4
Description of Some Simulation Tools

When this was written (January 1997), approximately 200 different simulation tools were available, each, naturally, with its own specific strengths and weaknesses. It would therefore be rather pointless to try and describe all of them. Moreover, this area of the software market is subject to enormous transformations, and one can roughly say that these tools undergo a "quantum leap" every two to three years.

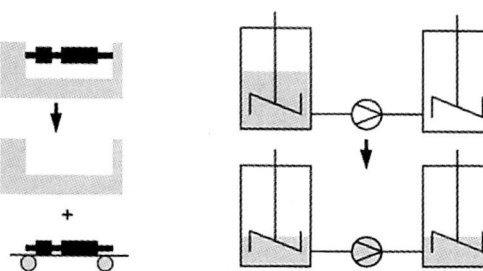

Fig. 9.4 Differences between the occupation characteristics of manufacturing technology and chemistry

In the following description, we have therefore concentrated only on the tools we have been using ourselves, or about which we know that they are suitable for dealing with process-engineering-type problems through personal contacts with other chemistry companies. For this reason, this can and should only be a subjective selection of what is available on the market, and therefore does not release future users from the responsibility of making their own assessments for deciding what simulation software is suited to their needs.

We have also limited the discussion to tools purely used for simulation. There are actually a large number of other production planning tools claimed to be capable of simulating material flow, but these are certainly not suitable for independent operation in process development or process revision. It is more likely that these tools have to be optimized for their later use with the aid of "real" simulators. More on such tools is found in Chapter 8 where production control is described.

SIMPLE++ [9.2]

SIMPLE++ is a standard software package for object-oriented, graphic, and integrated modeling, simulation, and animation of systems and business processes. SIMPLE++ is completely object-oriented: its graphic user surface, system architecture, and implementation are characteristic of object-orientation. SIMPLE++ is mainly used in production, logistics, and engineering, and there in almost all business branches as well as in research and development. It deals primarily with the optimization of structures, dimensions, and control of systems and business processes.

Chemical production systems can also be represented with SIMPLE++. For this, the software company offers a basic module that supports the creation of models. This module (SIMPLE++_process) assists the process developer with the optimization of recipes and the logistic design of production plants. Every model consists of production plant, recipes, and production tasks. Available for further refinement are resources, stores, breakdown-generators, and the use of stochastically divided operation times.

The production process as it progresses over time is schematically represented on screen (animation). During simulation, a large amount of information is gathered, which can be graphically represented at any time in the form of Gantt diagrams and graphs.

profiSEE [9.3]

The software package profiSEE is a specialized tool for the logistical simulation of production sequences in the processing industry, especially in the chemical–pharmaceutical branch. It has been developed by the "Operations Research Group" of Ciba-Geigy (Basel, Switzerland) to enable chemists and chemical- and process engineers to model batch and semicontinuous processes as well as multiproduct plants in a simple way and to analyze the corresponding material flows with regard to logistics (debottlenecking, production capacity). Numerous projects for optimizing monoplants and multiproduct plants have been carried out with the aid of profiSEE.

The main characteristics of profiSEE are the separation between plant and recipe, for transparent modeling, the user-friendly graphic surface, whose extended animation keeps the user continuously informed about the simulation process, and the compatibility with various types of hardware. Further properties worth mentioning are the exact consideration of campaigns, cleaning- and filter operations, and feedstocks, as well as the interfaces available to the widely used Office software packages.

BATCHES [9.4]

BATCHES is an American simulation software package of bpt (W. Lafayette, Indiana, USA). It combines the functions of an event-oriented material-flow simulator with simple thermodynamic modeling calculations. BATCHES has been developed to generate models and calculations of the various possibilities for cases where the process-engineering parameters have an influence on the dynamic course of the simulation.

When the model is produced, a strict separation is maintained between recipe procedures and plant structure. The recipe- and plant structure is produced graphically and interactively. The plant flow chart also contains, apart from the individual apparatus and their connections with each other, breakdowns and cleaning cycles.

The recipe consists of processes and process steps. A process usually summarizes all the process steps that take place in the apparatus or batch plant. The individual process steps are founded on thermodynamically based models.

During the simulation process, the program manages the vectors of extensive and intensive state variables of all the phases in the various apparatus. The simulation can be directed through any events that can be derived from the state variables.

POSES++ [9.5]

POSES++ supports, independently of the field, the development, simulation, and control of complex systems or processes that are characterized by discrete, discontinuous operation.

Special characteristics: client–server structure, high simulation speed, very large models can be developed, description of the models by a modular language, generation of specific surfaces for the simple execution of simulation experiments, graphic online animation, interface to the programming language C, possible integration into other software systems.

POSES++ can handle very complex problems. It has, for example, been applied in the process-engineering field for the simulation of product- and material flow through multilevel plants where the tasks could be described by parameters during planning. For the simulation of one production day, the computing time of a typical workstation is, at the moment, a single-digit number of minutes.

9.3
Modeling of Material Flow and Procedures

There are basically three questions that need to be dealt with regarding material flow in multiproduct plants:

- The dimensions of the buffers connecting the different process stages (batch, continuous)
- The dimensions of the resources, recognizing and avoiding resource "peak times"
- Determining the sequence of events for production

As in single-product plants, the connection of discontinuous or batch-oriented process elements and continuously operated units are physically represented by buffer elements such as storage tanks. In single-product plants, the size or capacity of these elements is determined primarily by the batch size. But the already mentioned dynamic behavior of the system of some plant parts can also here lead to unexpected material "peak times," which material-flow simulation should be able to deal with.

In multiproduct plants, the buffer capacity whose dimensions need to be determined also still depends on one further decisive variable: the sequence of events of production. This rather abstract and, where several products are involved, extremely complex concept is best explained with the following example, a problem that has been solved within the framework of a real investment project for a world-scale plant (approximately 200,000 t/year):

In a proposed new multiproduct plant, three products have to be produced. For this, the three intermediate products A, B, and C have to be processed further continuously in the centrifuges shown in Fig. 9.5. The centrifuges therefore also need to be continuously supplied by the accordingly filled storage tanks T 100, T 200, and T 300. The main stage of R 300 as well as the preliminary stages of both R 100 and R 200 each contains a discontinuous reaction. Not only do the reaction times differ from product to product and from stage to stage, but different quantities are also involved in the different stages. The continuously required quantities are approximately (not exactly!) in the proportion 1:1:3 (A:B:C).

The questions following from this are:

- What is the minimum size that the buffer vessels need to be so that production can proceed without availability problems?
- What would be a suitable production strategy for solving these problems?

At the fringe of this "buffer problem," a problem no less interesting, that of so-called dynamic resource balancing, appears:

- How much cooling capacity would such a production method with its partly exothermic reactions require?

The different products as well as the preliminary and main steps all, of course, have different cooling starting points and total cooling periods.

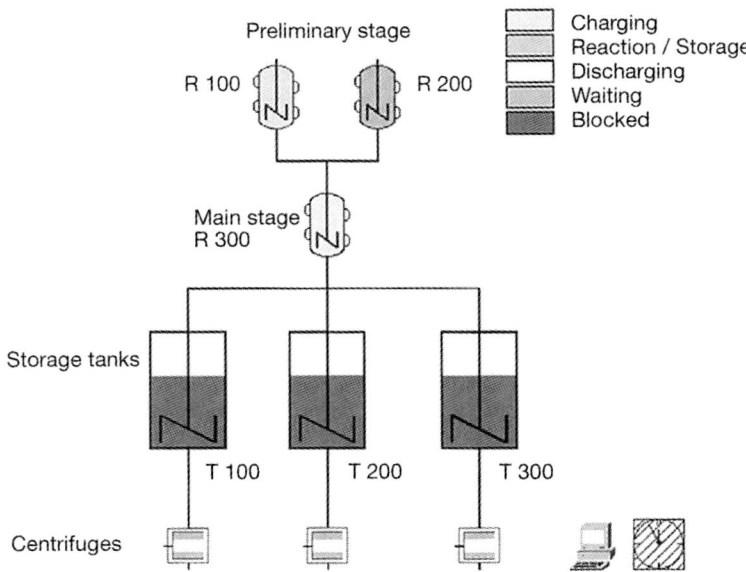

Fig. 9.5 Simulation model of a multiproduct plant with discontinuous and continuous elements

To solve these problems, the procedures and material flow for this production process were represented by the simulation program SIMPLE++. The original simulation surface is shown in Fig. 9.5.

With this model, various production strategies could be examined over a sufficiently long time period, to study the effect on the buffers and the required cooling output.

In Fig. 9.6, the changes in the storage levels of the three tanks are shown as the result of a simulation where production proceeds according to the basic pattern A, B, C, C, C (approximately according to the distribution of quantities mentioned above). The storage level at any time is determined by the balance of the tank content, actual inflow, and actual outflow. Since the outflow proceeds continuously and the inflow proceeds discontinuously in large quantities, the typical zigzag curve is found for the storage levels. If a curve touches or intercepts the zero line, that continuous production section needs to be switched off, leading to major problems during the later process steps. A 20-day period was simulated, which needed approximately 20 seconds of computing time.

It is clear to see that the chosen system A, B, C, C, C has deficiencies. There are already availability problems for product A after only three days and for product C after only four days. Component B, on the other hand, is produced in excess. On the basis of basic system A, B, C, C, C, a strategy was developed in which no availability problems occurred. The essential elements of this strategy are:

- A specific pattern (A, A, B, B, C, C, C, C) is followed at startup
- The basic pattern is adapted according to the state of the storage levels

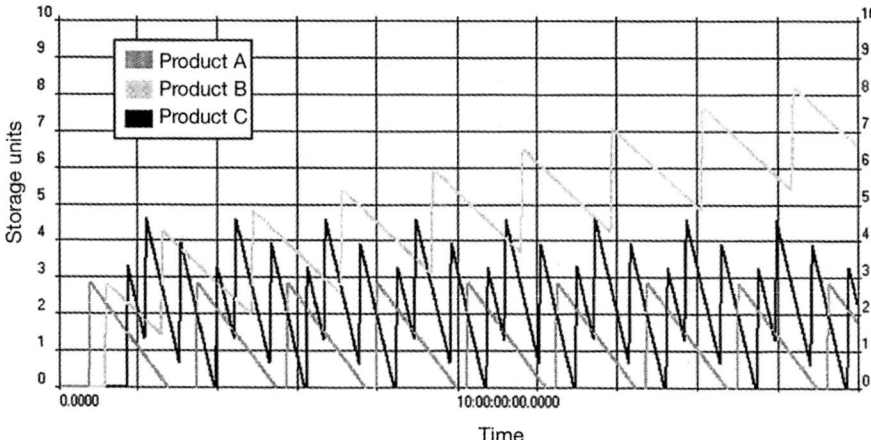

Fig. 9.6 Change in storage levels for basic type A, B, C, C, C; observation time, 20 days

In other words, during simulation, production proceeds as expected in the real-life situation.

The result shown in Fig. 9.7 convinced all the project participants. The specific, sequential loading of the storage containers allowed production to proceed according to basic pattern A, B, C, C, C. The first corrective measures to the production plan were only needed after approximately seven days. Instead of product B, product A was loaded, so that B did not increase excessively and A remained available for production. After a further nine days, component C was corrected, to prevent its storage levels from dropping to zero. The total required storage capacity found by this method was six storage units per product, three storage units less than determined by conventional ways, by methods not aided by simulation.

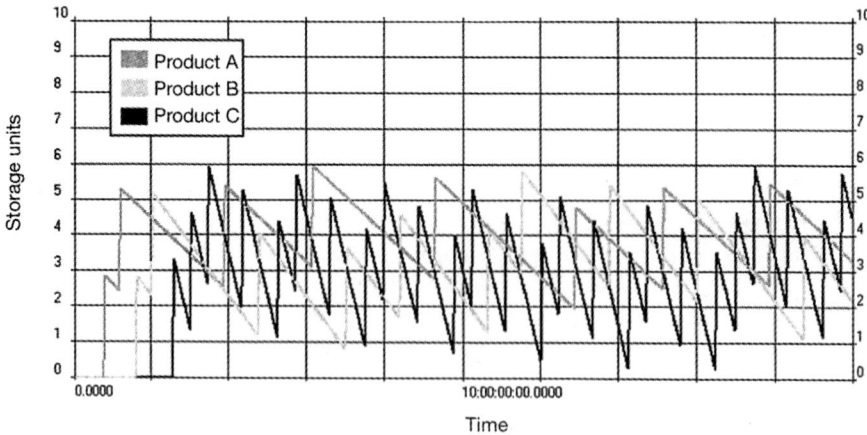

Fig. 9.7 Change in storage levels during "real" operation; observation time, 20 days

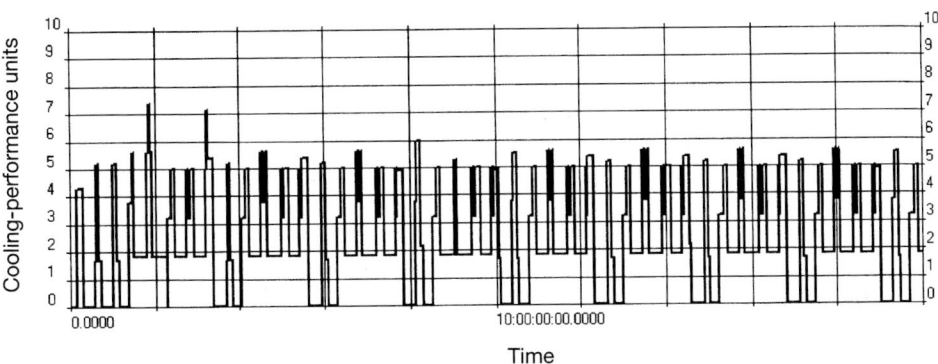

Fig. 9.8 Dynamic balancing of the necessary cooling performance; observation time, 20 days

The "byproduct" of this process was the dynamic balancing of the cooling performance. These results were no less interesting, as shown in Fig. 9.8. Peaks only appear at startup. A maximum cooling input of six units was enough to ensure the subsequent "stationary" operation.

This example shows the practical application of material-flow analysis in an investment project. In a similar way, this tool and, of course, the problem-solving methodology closely related to this tool can also be applied during process revision. Whenever one wants to evaluate the effects of specific measures, regardless of whether they are in the form of "iron and steel" or changes in the process sequences, the application of material-flow simulation makes sense. Or, to use a manufacturing-technology motto: *"simulation before investment."*

9.4
Economic Incentives for Material-Flow Analysis

The analysis of a multiproduct plant in the projection phase or for process- or plant revision is associated with costs that must be justified by the results. As this discipline is still relatively young in the chemical industry compared to the manufacturing industry, it is not yet possible to present adequate statistical data for the chemical industry.

Statistical evaluations of the benefits of dynamic material-flow simulation are available for the manufacturing industry:

- VDI guideline 3633: Savings 2–4% based on the investment costs at an expenditure of 0.5–1%
- FhG (Fraunhofer Association for Production and Logistics) simulation: Cost–benefit ratio better than 6:1 according to a survey of practical users

The cost–benefit ratio is the central issue in the projects known to us. Presently, the potential saving is especially high, because the benefit is calculated from the difference between the classical solution and the solution by simulation. In a few years' time, material-flow simulation will be state of the art, and every company will use it. Only the companies that start now will have an advantage over the competition.

Another important advantage, apart from the purely economical one, is that a better understanding of the plant system is gained. This is of considerable value, especially in the case of the very complex multiproduct plants.

9.5
Summary

The cases where these techniques have so far been applied in the chemical industry have shown that dynamic material-flow simulation is just as valuable in process engineering as it is in the manufacturing industry. Especially where a new multiproduct plant is being planned or the capacity of an existing plant needs to be expanded, a timely material-flow- and procedure analysis can ensure that the economic potential of the plant is realized to the full.

9.6
References

[9.1] Ensen, H., Fürer, U., Siebenhaar, W., Kurz, F., Kögel, G., *GVC Yearbook* **1995**, pp. 141–149.

[9.2] Company information, AESOP, Stuttgart (Germany).

[9.3] Company information, Ciba-Geigy, Basel (Switzerland).

[9.4] Company information, W. Lafayette, Indiana (USA).

[9.5] Company information, CPC, Chemnitz (Germany).

10
Plant Safety

10.1
Introduction

Chemical plants must be planned and operated in a way that ensures they pose no unacceptable risk to people and the environment. There should be no emissions that cannot be controlled safely or are not permitted by law. A plant is only safe if the pressure and temperature stay within the allowable limits, even during upsets. The safety of a chemical plant is determined by a number of factors:

- A safety-conscious plant design
- Faultless construction/assembly
- Plant maintenance
- Faultless operation

The technical discipline of plant safety should be a part of the planning and organizing of a plant design, which must consider the relevant safety issues. This design must deal with the potential problems and risks during construction, maintenance, and operation as far as possible, so that the risks associated with the plant is kept to a minimum. At BASF, a structured approach to safety analysis has become an established mechanism for achieving plant safety.

A plant safety plan consists of a number of components:

- Minimizing risk associated with the chemistry (materials/reactions)
- Minimizing risk associated with the process (pressures/temperatures/quantities)
- Suitable location
- Suitable equipment and materials
- Safe control (regulation/control/operability) of the process
- Safety-conscious planning details, including primary protection features
- Secondary protective features, that is, damage-limiting features for the "in spite of everything" case.

If the risks can be limited in the order given above at an early stage, the next stages become easier. The course must therefore be set in the "low-risk plant" direction at the earliest possible stage of the project. The characteristics of multiproduct plants (see Chapter 1) can soon lead to difficulties in meeting this requirement, as the products to

Plant safety plan			Location safety plan
Primary safety precautions		Secondary safety precautions	Emergency procedures
Avoiding the escape of materials		The "in spite of everything" case: the escape of materials	
		– with short-range implications	– with long-range implications
Control of the plant: "Operability"	Control of single errors with serious consequences: "Protective devices"	Controlling the escape of materials: "Limiting the damage"	Limiting the effect on the neighborhood
Control of: – Materials – Reactions – Processes During – Startup – Normal operation – Shutdown	Control of: – Technical failures – Product-related faults – Operating procedures – Surrounding influences, by means of: – Equipment and process technology – Process control technology – Organizational measures	– Fire protection – Fire-extinguishing equipment – Gas-alarm system – Isolation system – Rapid evacuation – Rapid pressure relief – Catch systems – Partitioning – Distance between structures – Measurement control room protection	– Risk-conscious location planning – Alarm systems – Evacuation plans – Neighborhood assistance – Availability and coordination of rescue systems

Fig. 10.1 Hierarchical subdivision of safety precautions

be produced in the plant may not even be known at the planning stage. Despite this, minimizing the risks in the order given above should always be kept in mind.

Safety precautions can be subdivided into four blocks (Fig. 10.1) [10.1]. The first three blocks concern the plant safety plan, which will be discussed in this chapter. The fourth block deals with the location safety plan.

The primary safety precautions should prevent escape of material from the equipment. Here *primary* means that it plays the most important role – the safest storage place for a dangerous substance is in the equipment, piping, or container intended and planned for it. Only in the "in spite of everything" case, that is, when these precautions have failed, should the "secondary" precautions come into play. They should ensure that the damage caused by the escaped materials is limited to the immediate vicinity of the plant.

Should that fail, and the danger spreads to a wider range – possibly crossing the chemical site's borders – then the emergency procedures developed for the chemical site as a whole need to be put in use.

It is impossible, within the context of this chapter, to address the specific precautions and to carry out a detailed analysis of the specific problems associated with multiproduct plants. This chapter has instead been structured so that a process in a specific plant is analyzed, the basic measures for it are introduced, and then, in conclusion, the specifics relating to multiproduct plants are looked at in more detail.

On the whole issue of plant safety it should be kept in mind that there are no *a priori* general rules for plants. It is much more important to analyze each specific case for the processes planned for that individual plant. At the end of this chapter, the reader should have an idea of how good planning of the plant program can keep the effort associated with the precautions to a minimum.

10.2
Structured Approach to Safety Considerations

10.2.1
Description of the Function

The minimum goal of a chemical plant is to produce a product of sufficient quality by an operation that runs with as few interruptions as possible, that is, with sufficient availability. That requires that the chemistry (materials and reactions) and the process (equipment and process control) can be managed reliably under all operating conditions (normal operation, startup, and shutdown).

As the first step of a safety analysis, the main hazards associated with the process should be determined or identified with the aid of the process description. The minimum measuring and control equipment ensuring trouble-free operation can then be defined according to these main hazard issues.

This "minimum solution" is usually significantly expanded, firstly due to various operating interests, such as:

- Product optimization: quality, yield
- Energy optimization: thermal integration
- Increasing the output: capacity reserves
- Increasing the flexibility: quantities, product types
- Increasing the availability: equipment in reserve
- Ease of repair
- Ease of operation.

10.2.2
Checking for Faults

Generally, additional features are also needed for plant safety. Therefore, the second logical step in a hazard analysis is to systematically check through the basic process technological plan, the "minimum solution," for safety weak points, and to deal with these through additional safety precautions.

Weaknesses or "faults" in the plant may be a result of:

- Technical failure of equipment parts or process control engineering equipment (e.g., failure of a regulator)
- Product-related faults
- Operating errors (e.g., entering an incorrect set point)
- Surrounding influences (e.g., frost).

10.2.3
Determining the Precautions

The following criteria should be used for a careful examination of the process-technological minimum solution:

- Where may faults arise?
- Do the faults have grave enough consequences to necessitate protective devices? If so:
- How should these be equipped to function properly?

Effective protective measures against faults with grave consequences could be any of the following ones:

- Precautions in the form of equipment or process technology
- Process control engineering units (see Chapter 7)
- Organizational measures

In Chapter 7, the procedure for determining the need for protective measures is represented by a flow chart.

Compared to control and monitoring units, protective units have to meet special requirements. They must function under all circumstances. The implementation of a process control engineering unit for protective purposes is described in greater detail in Chapter 7.

The logical route to answering the question of whether protective features are required or not is that of a risk evaluation procedure. Risk (R) is generally defined as the product of the extent of the damage S (consequences) and the damage frequency P (probability). Danger, in contrast, describes whether damage can result at all. It is therefore a nonquantified description of the damage potential. Risk is a quantitative description of the damage potential, determined from the probability of the cause occurring and the extent of the possible damage (Fig. 10.2).

For minimizing the risk, both the extent of the damage and the frequency of the cause can be reduced independently from each other by suitable modifications of the process and/or by means of specific measures. This must lead to the risk of operating the plant being lower than the acceptable residual risk. It is, however, not possible to determine acceptable objective and absolute quantities describing the risk of a plant or to determine the exact limits for the residual risk.

Whether, according to safety considerations, a precaution is necessary or not, or, to put it differently, whether a measure is sufficient or not, is determined by

Danger
Danger exists if there is a *possibility* of damage occurring
Danger is a *qualitative* concept

Risk
$R \equiv P \cdot S$
where:

R = risk
P = damage frequency (probability of damage occurring)
S = extent of the damage (effect, consequences)
Risk is a *quantitative or qualitative* concept

Fig. 10.2 The concepts *danger* and *risk*

whether it is possible for a single error to lead to damage to the plant that may further lead to grave consequences for people and the environment. For this, a systematic analysis of the process is necessary, so that all the main risks are recognized, and so that the precautions can be applied to a sufficient extent and in the correct situations.

The simultaneous occurrence of several errors, independent from each other, is usually not considered, as the probability is negligible. It is, however, essential to determine exactly how independent the possible errors are from each other and whether or not it is possible for one fault to be undetected for a long time (because it does not detract from the operation) until another fault occurs. In such cases, it should be clarified for the sake of plant safety if, for example, a temperature switch that is not regularly used during normal operation does not function exactly when it is needed most, when another fault necessitates its functioning. Another example is faulty charging during a semibatch process; this may lead to an increased accumulation of reactive materials, which can further lead to a runaway reaction. If the cooling functions without trouble, and enough reserves are available, no significant disadvantages to normal operation would occur; this charging fault could then remain undetected for a long time. Should further faults occur, namely breakdown of the cooling system, this can have considerable repercussions, which would not have been the case if charging proceeded fault-free.

The simultaneous occurrence of several apparently independent faults must, however, be considered if the faults have a common cause ("common-mode" faults). For example, the breakdown of two independent cooling devices must be considered if the cooling medium is supplied by the same supply pump. It should also be asked, for example, what would happen if the nitrogen supply breaks down, so that neither an inert atmosphere nor the gas buffer operating the instruments is available.

10.3
Safety Considerations for a Process in an Actual Plant

If a process should be carried out with due regard to safety, the following points must be considered during planning and for operation, and must be backed up with sufficient data so that the danger potential can be evaluated:

- Physical properties of materials
- Chemical reactions
- Location and building plan
- Process plan and construction form

From the point of view of process planning, the following points should be considered:

- Reverse of materials
- Escape of material from industrial discharge points
- Leaks in plant parts
- Process control engineering plan for plant safety
- Environmental damage

Different keywords can be used for working through and discussing the technical safety problems point by point. As with the guidelines for the HAZOP (**HAZ**ard and **OP**erability) [10.2] or PAAG ("**P**rognose von Störungen/**A**uffinden der Ursachen/**A**bschätzen der Auswirkungen/**G**egenmaßnahmen", prognosis of faults/finding the causes/estimating their consequences/countermeasures) studies [10.3], it has proven useful to first define the desired values of a plant unit and then to discuss deviations from these set values with the use of guidewords [e. g., *no or not (no or none), more, less, and also, partly, opposite to, different to*], where the depth and detail of the analysis and documentation should be adapted to the problem at hand.

In the appendix to this chapter, keywords for different significant subjects are listed. This list is not comprehensive; it is merely intended as an aid for considering or mentioning as many of the relevant issues as possible. Complete sets of data need not be generated on all the keywords for each particular problem analysis; the decision for omitting a detailed analysis should, however, be justified and documented.

10.3.1
Taking the Properties of Materials into Account

Physical data, chemical properties, characteristic safety variables/properties, and toxicological and ecotoxicological data all fall under the heading of substance properties. The toxicological data aid the discussion on the industrial safety measures. The ecotoxicological data (e. g., water-pollution class, bioaccumulation) are important for determining the measures needed for environmental protection, especially for the prevention of water pollution.

The combined substance data is the most important basis for the safety-conscious design of the planned equipment. To realize the aim of a "tight plant, from which nothing escapes," the corrosion data are of primary importance, so that the correct materials are chosen.

Once the physical properties of the materials are known, their handling can be discussed. Here one distinguishes between materials supplied through pipelines as opposed to those supplied in containers. The issues of toxicity and protection against explosions determine the sealing quality required of the connections between the equipment. What also needs to be dealt with in this context is whether the materials should be handled in completely closed systems, or whether more or less open handling, for example, emptying of bags, is permissible.

Other examples such as room-dust monitoring, technical ventilation, and the use of encapsulated pumps, special valves, or steam and water curtains are additional industrial safety measures adapted to specific material properties. With some materials it may even be unacceptable to release safety valves directly into the atmosphere. In such cases, special measures, such as compartmentalization, inherently safe pressure designs, or comprehensive disposal measures are necessary.

10.3.2
Control of Chemical Reactions

After the properties of the materials that are used for and formed during the process have been considered, the control of the chemical reactions need to be ensured if a chemical transformation is taking place. Such problems are dealt with in publications of, for example, the German Technical Committee for Plant Safety (TAA, Technischer Ausschuss für Anlagensicherheit) and the organization BG Chemie [10.4, 10.5]. As basis for the discussion, the stoichiometry, the possibility of gas being generated, the reaction conditions, the mass balance, the yield, and the formation of side and intermediate products need to be exactly characterized.

Knowledge of the following variables is also essential for a comprehensive discussion:

- Thermokinetics: The heat release associated with the reaction should be known, including its integral value, and its development over time. This provides the information on the behavior of thermal accumulation, that is, the accumulation and subsequent consumption of the not yet reacted starting material in the reaction mixture as the reaction proceeds.
- Thermal stability: With microanalytical (e.g., DSC, differential scanning calorimetry measurements) and adiabatic test methods, the thermal stability of the reaction mixture needs to be determined to ascertain safe working methods. Following this, questions on the behavior of the system when normal operation is deviated from (e.g., cooling faults, no mixing, dosing errors, etc.) need to be posed and answered.

10.3.3
Choice of Location and Construction Design

Once the danger potential associated with handling the materials and carrying out the reaction is known, the location of the intended plant needs to be evaluated. For this, the possible effects of the plant on the neighborhood and how the neighborhood may affect the plant need to be clarified. On the basis of the discussion on the location, issues related to logistics and product changes can then also be evaluated.

With regard to the construction design, issues such as the building plan, the ventilation of the buildings, the accessibility to the operating staff and the fire brigade, the number, location, and design of the emergency routes, the draining of the work surfaces, the collection and disposal of leaks and fire extinguishing water, and the collection of product or operating materials during maintenance need to be examined.

Finally, drawing up a suitable fire-protection and fire-fighting plan is also one of the tasks that need to be carried out during the safety planning of a process.

10.3.4
Process Design and Construction

10.3.4.1 Inherently Safe Processes
Guided by a knowledge of the danger potentials, one can now develop the process plan and its constructive form. During process development, the possibility of an inherently safe process, that is, a process that is safe in itself, should first be investigated. Instead of limiting, to the best of one's ability, the recognized dangerous instances by a number of safety mechanisms, one should rather first try to remove or avoid the dangers altogether. According to Kletz [10.6], the following principles should be followed if an inherently safe process is to be realized:

- Lessening/minimizing
- Replacing/substituting
- Reducing/moderating

For the logical planning and setup of an inherently safe process, the following points should be considered:

- Simplifying
- Changing early on

The principle of "lessening" means a decrease in the amount of dangerous materials being held up. For example, a continuous or semicontinuous (semibatch) process is preferable to a batch process; distillation units should be supplied with very efficient packing, so that the amount of dangerous materials at the bottom of the column is lessened. Dilution is also a possibility worth exploring.

The principle of "replacing" means that dangerous materials should, wherever possible, be replaced by less dangerous ones or modified. For example, for the purification of waste water, chlorine should be replaced by sodium hypochlorite, so that the dangerous transport and storage of liquid chlorine is avoided. Flammable or toxic solvents should be replaced. If, during synthesis, a critical intermediate is formed, one should first check whether an alternative process, in which no dangerous intermediates form, is available. If another process cannot be found, it should be attempted to ensure that the intermediate is transformed immediately after its creation. This way, the transport and storage of the dangerous intermediate can at least be avoided.

The principle of "reducing" means that handling at high pressures and temperatures should be avoided. For example, a liquefied gas should be stored at low temperature and low pressure, instead of at ambient temperature and the associated high pressure. In case of a leak, the gas cloud resulting from the gas stored at low temperature is smaller, and emergency measures can be applied more effectively. A catalyst can also facilitate the reduction of pressure and temperature during reactions.

The principle of "simplifying" embodies the aim of avoiding additional safety devices by getting rid of the dangerous instances; this makes it easier to grasp the setup of the plant and thus makes it safer. Inherent safety must be the goal of every planning session. And if something has to be changed, it should be done early on. For example, combustion of solvent vapors should, to avoid emissions, be replaced by low-temperature condensation: this is a safer process, as there is no danger of possible flashback from the combustion unit.

Finally, where possible and economically viable, pressure-resistant or pressure-surge-resistant equipment should be chosen. This may not lead to "inherently safe" processes as defined by Kletz, but such processes can at least be regarded as inherently safer, since the equipment is built to withstand the expected pressure increases without being damaged. A chemical plant is therefore inherently safer if the protective goal – here more or less the prevention of an unacceptable overpressure – is achieved without the aid of process control engineering units, other technological safety measures, and without the intervention of the plant personnel.

It should be remembered that it is difficult to achieve an absolutely inherently safe process. It is more a case that specific process variables can be made inherently *safer*. When a change is undertaken to minimize the hazard potential, it should be checked carefully to ensure that this change does not lead to other new or increased hazard potentials. For example, a low-boiling solvent may be used to limit the temperature of an exothermic reaction; however, it may be that the regular operating process temperature is already above the flash point of the reaction mixture, thus causing an explosive atmosphere.

10.3.4.2 **Examples of Questions for Determining the Dangerous Instances in a Process**

When the employed equipment is being safeguarded against unacceptable temperatures and pressures outside the design limits, the keywords given in the appendix of this chapter should be consulted for possible causes for the design limits being exceeded.

Depending on the type of pump used, special precautions against blockage, running dry, and reversed operation should be taken.

For heating with organic heat-transfer media, the special instructions from BG Chemie [10.7] should be followed. But apart from following these instructions, one should consider whether the maximum temperature values of the heating/cooling medium is suited to the thermal stability of the material for whose heating it is used. A temperature which is too high should be avoided if the material already visibly decomposes in the accessible temperature range. Temperatures that are too low should also be avoided if materials with high melting points are around. The heating/cooling medium should also not be reactive towards the process media (internal leakage in heat exchangers).

The plant and process should be designed in a way that precludes undue vibrations in the system, for example, due to machines. Attention should also be paid to certain flow phenomena, such as abrasion, erosion, cavitation, dead-volume generation, and so forth.

Temperature changes that are too rapid should be avoided, otherwise damage to the mechanical structures may result; this may lead to the release of materials. For the same reason, material fatigue should reckoned with.

The option of keeping the operational contents confined should be considered carefully to avoid problems resulting from, for example, contained liquids expanding under the influence of external heating. In continuous plants, it is generally unavoidable that certain plant sections need to be isolated, because in this way the quantities in specific plant sections can be reduced in case of a mishap.

The process design should be checked continuously to ensure that all operating conditions can be coped with. The following points should be clarified: startup and shutdown procedures; deviations and cross-contamination during product changes; rapid switching on/off; agitator setup (heat transfer, mixing behavior, foaming); extent to which equipment is filled; functioning ability of the agitators; temperature measurement and heating/cooling circulation under all operating conditions; practicality of the respective connections of the equipment.

The reverse problem both within the plant as well as within the site network and/or the complex should be considered. The relevant keywords and themes are compiled in the appendix.

A further, important consideration during process design is how to handle materials escaping from the industrial outlets. In this context, the danger posed by the escaped toxic materials or the development of explosive air/fuel mixtures should be discussed.

Leakage of materials is another group of themes that should be analyzed, and then not only with regard to the individual plant parts, but also by tracing its possible paths, from the outside to the inside and vice versa, and then in relation to the environment as well as other equipment. The question of self-ignition phenomena upon leakage into isolation should also be considered. Apart from secure leakage detection, plans for the control of leaks (confinement, collection, disposal) should also be developed.

A strategy for coping with possible power loss or auxiliary power failures should also be developed. Following a power loss, automated valves should switch, powered by springs, to a predetermined fail-safe setting (open or closed). In this way, the system remains in, or achieves, a safe state.

Finally, the question of explosion protection should be dealt with. According to the literature [10.8, 10.9], an explosion classification should be undertaken. The possibility of explosive gas or vapor and air mixtures developing in the individual plant units and their spatial environment should be determined so that the necessary precautions can be taken if necessary: primary explosion protection, that is, the prevention of explosive mixtures, and secondary explosion protection, the prevention of sources of ignition.

The process control engineering concept of the plant should be checked from the point of view of plant safety especially to determine which of the already planned or installed and additional process control engineering units have a protective function (see also Chapter 7 on this). Special demands, defined in the guidelines (see Chapter 7), are placed on such protective facilities. The difficulty in determining the protective function lies in having to exactly determine which primary cause can lead to faults with dangerous consequences and to analyze whether the faults or events leading to faults can or should be interrupted simply by a process control engineering measure. If this is the case, the process control engineering protective device is designated as being in Class A.

10.4
Special Features of Plant Safety in Multiproduct Plants

In this section, the special features of determining a safety concept for multiproduct plants is discussed. For this, it is assumed that some of the key themes from the previous chapters have been considered during the basic planning of the multiproduct plant, for example, that the pumps should not cause the equipment to be placed under excessive pressure or that the flushing of a column should not lead to any mechanical damage.

After the hazard potentials in a plant planned and built for a specific process and product (a so-called monoplant) have been recognized and assessed, tailor-made measures can be taken to ensure safe operation. In multiproduct plants, these tailor-made measures cannot apply to all the processes run in the plant, as this would compromise the flexibility of the multiproduct plant, and therefore its characteristic nature and advantage.

A multiproduct plant is characterized by, amongst other things, the process control engineering and energy plans of the plant that are designed to make extensive variations of several process parameters possible. As a rule, the desired total variation range exceeds the technically safe limits allowed for the individual processes. This means that a multiproduct plant is supplied with variable switching units, which limit the operating conditions for each specific product to within the technically safe region, the so-called window. With this requirement that the limiting values should also be individually and routinely adaptable, the danger of human error does, however, once again come into play.

10.4.1
Examples to Clarify the Nature of the Problem

The following examples illustrate the problems associated with determining a safety program/concept for a multiproduct plant. Where the examples do not deal directly with possible solutions for multiproduct plants, possible solutions are shown in Section 10.4.2.

Example 1
The heating media used in a multiproduct plant allow a maximum temperature of 200 °C. A solid whose decomposition reaction rate is already significant at 150 °C is to be treated in the reactor. This means that if a large amount of solid bakes onto the reactor walls above the liquid level, decomposition is unavoidable with heating that is too aggressive or in the absence of temperature-limiting features. In a plant tailor-made for this process, a heating medium whose maximum temperature is below the allowed maximum would have been used. In a multiproduct plant, however, the maximum allowed temperature would need to be held by a switch specifically designed for this product. A special risk evaluation is then also necessary to determine the consequences of an erroneous classification or setting or the technical failure of such a device.

Example 2
For some of the products, the heating medium can, even at its maximum temperature, not reach vapor pressures exceeding the pressures planned for the plant. For a product introduced at a later stage, this relationship may no longer be valid. The maximum temperature of the heating medium may even lead to unacceptable situations due to the vapor pressure. For this reason, it is generally not possible to reach the level of inherent safety defined by the vapor pressure curve of the feedstock with regard to the obtainable pressures in a multiproduct plant.

Example 3
In a multiproduct plant various products are manufactured for which, during the reaction at temperatures below 100 °C, water has to be added. Another product needs to be produced at 150 °C. Here water addition at temperatures above 100 °C is completely out of the question, as this will certainly lead to the resulting water vapor putting excessively high demands on the pressure-relieving lines, and leading to equipment rupture in extreme cases. In tailor-made equipment, this water connection would not have existed at all.

Example 4
In a semibatch process, a reactant needs to be added within a specific temperature window. At too low temperatures, the reaction becomes sluggish, and the reaction components accumulate; should this reaction mixture then reach a higher

temperature (e.g., during later stirring), heat may then start being generated at an uncontrollable rate. This is also already a case where only one error, namely the disallowed deviation from the reaction temperature during charging, can lead to unacceptable consequences. For this reason, a tailor-made plant would have a temperature switch as a protective device for dosing. Such a switch would ensure secure closing of the inlets at temperatures outside the permitted region.

If the safety of a process depends on the functioning of only one specific switch, this switch is categorized as a class A protective device according to the NAMUR classification system. Then its technical design and operating principles are subject to strict rules. If the switch has responded, it can not be allowed to automatically reset. The operating personnel may not remove or open the locking mechanism of the inlet. The switch set points may also only be set or altered by the technical department in charge and all changes to this switch must be documented.

If a new material is produced at temperatures outside the current set limits, it is easy to see that the heavy restrictions outlined above will result in complicated organizational procedures. For this reason, the design of processes in multiproduct plants that do not need class A switches is generally aimed for, and other technical safety measures for achieving the protective function (e.g., avoiding the escape of materials) are looked for.

Examples 1 to 4 illustrate that it is already possible for unacceptable situations to arise due to simple errors, such as when a new process in a multiproduct plant has a reaction system and recipe that cannot be accommodated by the current set-up of that plant with regard to equipment, technical processes, and process control technology.

One should always be aware that the effort associated with planning technical safety aspects, which before the building of the plant was the responsibility of the planner, will at a later stage be transferred to the operator of the plant.

10.4.2
Guaranteeing the Safe Operation of a Multiproduct Plant

How does one practically go about guaranteeing safe operation? As already mentioned before, there are basically three different blocks of measures to ensure safe running of a plant:

- Measures related to process engineering and equipment
- Process control engineering measures
- Organizational measures

Measures Related to Process Engineering and Equipment
When planning a multiproduct plant, one first needs to ensure that the plant operates as designed, that is, "operability" should be ensured. One should try to cover the expected range of variations in operating parameters as far as possible. That means that the equipment should be designed to cope with the highest ex-

pected operating pressures, temperatures, and filling levels (load), and to meet the highest required standards regarding explosion protection, anti-corrosion measures, and so forth. With regard to pressure, the entire facility should be designed uniformly (vessel, column, cooling, etc.). Even if these requirements are affordable, the plant is not necessarily safe yet, because deviations from planned operation and the control of such undesired situations still need to be considered. The control proceeds by, among other things, process control engineering measures.

Process Control Engineering Measures

To ensure safe operation by means of process control engineering devices, one must first, as already described before, ascertain whether the process control engineering unit plays a protective role and therefore needs to be classified as a class A facility or not. Besides this fundamental decision, with multiproduct plants one always needs to reconfirm whether the process control engineering protective unit that has been defined for a specific process can at all guarantee the protection of another newly introduced process.

An example of this is the charging of a component with high vapor pressure. Here, pressure measurement and switching off at high pressure is suitable for recognizing when a large quantity of the charged component has accumulated, so that additional charging can be limited by the inlet being closed off. This concept is not suitable for detecting components with very low vapor pressures.

If a process control engineering unit has been defined as a protective facility, its design should conform to the NAMUR recommendations.

The process control engineering switches and regulators should be discussed and determined for every newly introduced product by a group of experts following a strict set procedure. The local changes should be monitored and documented. It is a good idea to test the switches and regulators with a "water run" before the first run of a new product.

If only one specific class of compounds is to be produced in a multiproduct plant, it may then be possible to limit the organizational effort associated with changing the set points for a specific recipe (of the same product class) by the use of safety-focused programmed circuitry for the set points important for technical safety. When safety-focused programmed circuitry is used in conjunction with a class A process control engineering unit, it must be unambiguously clarified whether this latter function is guaranteed in combination with the programmed circuitry. In addition, it should always be checked again whether the process control engineering unit that is being used is capable of guaranteeing the protective goal for all the recipes of the relevant product class. If products belonging to different product classes are to be produced in a multiproduct plant, the use of programmed circuitry is probably not a good idea.

Organizational Measures

In multiproduct plants, organizational measures make up a substantial part of the safety program. This can be illustrated with the examples given in Section 10.4.2

for newly introduced products. For the heating medium, for example, the following organizational measures could be taken: where vapor heating is used, an additional pressure reducer may be temporarily employed. To prevent inadmissible addition to the reactor, the inlet may be temporarily disconnected from the unit with blind caps, rotating curved pieces, or adapters.

If dosing should proceed within a defined temperature window, then – as long as no class A switches are in use – detailed operating instructions should be prepared and demonstrated, so that the operating personnel can follow the procedures safely and there is no haphazard adjustment of set points. In this context, it should be checked whether other safety facilities, such as safety valves for limiting the outcome of an "in spite of everything" case, are required or not. Then the organizational measures, safeguarding the correct adjustment of the set points, are aimed at preventing the safety valve from responding. The protective goal, which may, for example, be that of preventing the equipment from rupturing, is then taken over by the pressure-relief valve.

The organizational aspect becomes especially obvious in the context of the cleaning procedures necessitated by the frequent product changes. It must be ensured that there are no undesirable interactions between the impurities and the cleaning agents. The detergents used for one process may have negative effects on another process. Here one should also, where necessary, make use of blinds, adapters, or similar features to prevent specific materials from being added at the inappropriate moment or being around at the same time.

Apart from the organizational side, the construction aspect is also important for cleaning procedures. It must be ensured that product residues can be removed completely from the equipment and lines and that there are no dead spaces that cannot be evacuated (for example, by use of higher-lying evacuation connections). Otherwise, products may collect in these dead spaces and undergo undesired reactions with materials from other processes. The evacuation ports must be protected against freezing. The number of possible sources of errors can be reduced by a clear pipeline layout and appropriate labeling of the pipes.

Where hoses are necessary, care should be taken to ensure, for example, that the hoses are suited to the transported media and also that electrostatic charging is safeguarded against.

In systems such as waste-gas and waste-water facilities that are shared by various plant sections, it should be ensured that the materials in question are not incompatible, that there is no back flow into other plant sections, and that the materials cannot accumulate.

For the organizational measures to function properly, the plant needs to operate within an environment for which the necessary preconditions have been put in place. Parts of these preconditions are specific planning and control mechanisms determined and demanded by management.

"Change is the only constant" is a very fitting motto for multiproduct plants. Therefore management of change plays a very important role in multiproduct plants, which are characterized by frequent production changes. In multiproduct plants, a comprehensive safety analysis by a multidisciplinary team should be an automatic procedure before the introduction of a new production process. To aid this procedure, a transfer protocol, in which the characteristics of the new production process are recorded, should be drawn up, to form the basis of the discussion.

Even for small changes, for example, the production of a higher homologue of a product family, a mechanism should come into place in which the technical safety issues are discussed by a small but competent group. Such a discussion should not only occur for newly introduced production processes, but also if the plant is altered, even when the product synthesis remains unchanged, so that the repercussions of these changes on other product syntheses can be analyzed.

During the safety analysis, the critical points for each production process should be worked out and the special organizational measures should be discussed and decided upon. These measures should be checked for their effects and may not be changed arbitrarily.

The organizational measures decided upon should next be consistently implemented in the plant. Management should ensure that the necessary preconditions in the plant are in place so that the measures can be passed on to the operating personnel in full detail. It must be ensured that the sense and the purpose of the measures are understood by the staff. There should, for example, be an operation-management representative with enough time dedicated to safety issues. The transfer and application of the measures need to be checked at sufficient time intervals. In a multiproduct plant, personnel who are particularly well-trained and suited to the task are required in sufficient number to ensure smooth running.

It has also proved useful to introduce batch charts for the individual production processes, as proper operation is ensured by the staff using these, and the stipulated course of the process can be followed. The batch charts should contain specific points of reference against which the proper course of the process can be verified and for which specific measures are given in case of deviation.

Wherever possible, the production prescription and other instructions should be incorporated as a procedure control into the process control system. In this way, human error can be reduced to a minimum. Software should, however, not be overestimated as a cure-all. Big and complex program structures are exactly the situations that provide software with enough error possibilities for causing critical situations in the plant. The application of software should therefore, for safety reasons, always be closely observed and critically regarded.

Clustering specific plant parts or reactors needed for specific product families or syntheses is, where feasible, a very useful way of keeping organizational measures within bounds. This may also minimize the effort associated with product changeovers. Where exactly defined connection changes are required during this work, the use of adapters is preferable. With complicated connection changes, the technical safety consequences can, for example, be highlighted by use of the PAAG process.

If plant parts have been clustered together and only specific products are run in these plant sections, the precautions can be adapted to the most critical case. The set points or, for example, the maximum allowed dosage rates, are determined for the most critical cases. As long as the plant has not been altered and the modified plant has undergone no additional technical safety checks, one may then, in this plant section, only produce products whose relevant technical safety characteristics are less critical than those used for the plant design.

This type of procedure is also recommended for the application for official approval. The documentation for permission from the authorities can only be directed at the most critical situation. On the basis of this, one can then aim to reach an agreement with the authorities by which it is accepted that no additional approval is required if the products manufactured, also new ones not contained in the application, do not have more critical technical safety characteristics. This procedure can be regarded as analogous to the procedure related to explosion protection. In this case, the product with the lowest ignition temperature determines the choice of the electrical equipment. New products with higher ignition temperatures do not necessitate a new examination and classification.

For the different multiproduct plant versions mentioned in the introduction – discontinuously operating multiproduct plants, continuously operating multiproduct plants, modular multiproduct plants, multiproduct plants with pipeline manifolds, pipeless plants, and multipurpose equipment – the above systematic procedure for identifying and assessing danger sources can be applied as with monoplants. There are, however, issues specific to multiproduct plants that need more attention.

In discontinuously operating multiproduct plants, modular multiproduct plants, multiproduct plants with pipeline manifolds, and pipeless plants, the greater chance of confusion should be reduced to a minimum. For example, no unsuitable plant parts should be combined for the corresponding tasks; no incompatible materials should come into contact with each other; no materials may enter plant sections that are operated with process media whose process parameters (e. g., temperature) lie outside the safe handling limits of that material.

The materials handled in a multiproduct plant should have technical safety characteristic data that allow their use within the limited conditions of that multiproduct plant. For example, with multipurpose equipment such as driers, the explosion dangers need to be discussed in detail. Is the atmosphere possibly an explosive one? Is there a likelihood of ignition sources being available?

For pipeless plants, in particular, the danger of undesirable pressure buildup should be discussed, since the normal procedure with pipeless plants involves the transportation of closed containers during production. Is a pressure-release device available for dealing with excess pressure, and, if so, how is the released material handled?

10.5
Function and Design of Pressure-Relieving Equipment

Despite all efforts beforehand and during the introduction of new production processes, errors cannot be excluded completely. For this reason, it is essential that the equipment in use is safeguarded against bursting as a result of too high disallowed pressures. This can be achieved by pressure-resistant or pressure-surge-resistant construction or safety valves and/or rupture disks.

Safeguarding by means of safety valves therefore always proves to be the last resort. The safety valve has to be designed for a specific situation, and it is recommended that the most critical product is chosen for planning the design. Every new production recipe should thereafter be checked to ensure that the design criteria of the safety valve and the subsequent pressure-release facilities can accommodate a runaway reaction. Should this not be the case, it is recommended that the production prescription is changed so that the dimensions of the safety valve can deal with the worst case of a runaway reaction, also for the new recipe.

But simply redimensioning the pressure-relief unit is not sufficient for dealing with the entire problem. The safe handling of the mass flow resulting from pressure relief needs to be considered too. The problems related to pressure release are dealt with in reference [10.10], which gives recommendations specific to the nature of the various materials being released.

10.6
Summary

From the above discussion on the specific problems related to multiproduct plants, it becomes obvious that of the three possible measures for ensuring safe operation, namely:

- Measures related to process engineering and equipment,
- Process control engineering measures, and
- Organizational measures,

the organizational measures play a very important role in the safety plan. For this reason, it will be pointed out here once again that, for the operation, an environment needs to be created in which the organizational measures actually function properly, where no problems that can lead to negative events develop in the interfaces human–human and human–technology.

In the comments above, the prime concern was with production in general, which also consists, apart from the actual synthesis, of all the technical operations associated with the process, such as drying, grinding, distillation, or formulation. What has been discussed above applies equally to these.

10.7
Appendix

Keywords on different key subjects
(This list is not claimed to be comprehensive!)

Properties of Materials: Physical Data

- Melting point
- Boiling point
- Vapor-pressure curve
- Solubility (also of gases)
- Sublimation temperature
- Transformation enthalpies
- Electrical conductivity
- Electrostatic chargeability
- Viscosity
- Density
- Molecular weight
- Critical data
- Thermodynamic phase diagrams
- Heat capacity
- Conductivity of heat, etc.

Properties of Materials: Chemical Properties

- Self-reaction capability of the material (e.g., polymerization, catalytic decomposition)
- Peroxide generation
- Reactivity towards heating/cooling medium, towards iron (or other construction materials) or towards sealing materials (such as gaskets, etc.)
- Corrosive properties
- Exothermic nature and kinetics of self-reaction

- **Properties of Materials: Characteristics Pertaining to Safety**

Liquids and Solids

- Flash point
- Ignition temperature
- Flammability
- Self-ignitability
- Self-ignition temperature
- Deflagration ability
- Thermal stability (for determination of temperature limits)

Dusts

- Dust-explosiveness
- Minimum ignition energy
- Ignition temperature
- Possibility of a hybrid mixture forming

Gases/Vapors

- Explosion limits
- Minimum ignition energy

Handling of Materials – Products Available through Pipelines
- Mechanical damage
- Reverse flow (safeguarding of network)
- Freezing or disallowed heating
- Pressure protection
- Influence of pressure and concentration fluctuations
- Thermal expansion of liquids when accompanying heating is used (enclosure of media)
- The Joukowski phenomenon in very long lines during rapid closing of valves, etc.

Handling of Materials – Product Available in Containers
- Suitability (electrostatics, etc.)
- Disposal of packaging
- Electrostatic charging during filling, dosing, and transport
- State-of-art design of filling stations
- Safeguarding against overfilling and mixing-up of products
- Protective measures, such as grounding, inertization, against explosions
- Operating instructions in the form of a checklist

Keywords on Chemical Reactions
- Presence of a multistep reaction
- Critical intermediates are concentrated (enriched) under specific conditions
- Occurrence of undesired self-acceleration or autocatalysis
- Specific starting conditions required for a reaction regarding composition, temperature, or pressure
- A reaction becoming dormant/recognizing a reaction taking off
- Possibilities for quenching a reaction

Response of the Chemical Reaction to Deviations from Normal Operation
- Response to incorrect composition of the reaction mixture (too much, too little, none, other components)
- Sensitivity of the system to impurities
- Response to incorrect charging of a component (with regard to type, quantity, sequence, rate)
- Interactions with the auxiliary agents
- Behavior at a too high or too low reaction temperature or a too high or too low pressure or the incorrect pH value
- Heat removal is hindered and/or the reaction breaks off as a result of viscosity changes or the development of phase boundaries

Causes for Pressures Exceeding the Planned Limits
- Back flow (especially from adjacent equipment)
- Leaks (damage to equipment/pipelines with pressure drops)
- Inlet/outlet flow quantity; internal or external heating

- Reaction (desired and undesired)
- Ignition of explosive mixtures
- Power outage
- Operating error
- Fire
- Flushing (overfilling)
- Evacuation through suction or siphoning

Causes for Back Flow
- Breakdown of delivery unit (pump, blower, vacuum unit)
- Pressure rise in plant section due to uncontrolled reaction, impurity of manifolds, mechanical damage
- Siphon evacuation due to too low pressure
- Overfilling of equipment
- Power outage
- Blockage
- Closed manually operated valves
- Shutting possibilities in composite system, operating error

Keywords on the General Subject of Industrial Waste Discharge
- Pressure-relief devices (safety valve, rupture disks)
- Permissibility of the release
- Design criteria for the dimensions of the relief unit
- Diffusion calculation
- Disposal facilities (scrubber columns, combustion units)
- Gas erupting in liquid-phase separators, at the bottom of columns, siphons, etc.
- Bursting through of a phase during liquid-phase separations
- Stress-relief line of a double barrier
- Sampling devices
- Discharge and rinsing connections
- Vent and rinsing lines of process-control-engineering facilities

10.8
References

[10.1] KÖSTER, H., *Grundkonzept Anlagensicherheit [Basic program for plant safety]*, Internal paper of BASF AG.

[10.2] Chemical Industries Association, *A guide to HAZARD and OPERABILITY STUDIES*. Publications Department, Chemical Industry Safety and Health Council of the Chemical Industries Association, Alembic House, 93 Albert Embankment, London SE1 7TU, Great Britain (and references therein).

[10.3] Internationale Vereinigung für Soziale Sicherheit [International Association for Social Security], Berufsgenossenschaften der chemische Industrie [Professional Association for Safety in the Chemical Industry]; Gaisbergstr. 11, Heidelberg, Germany.

[10.4] Technischer Ausschuss für Anlagensicherheit (TAA) [Technical Committee for Plant Safety], *Erkennen und Beherrschen exothermer chemischer Reaktionen [Recognizing and controlling exothermic chemical reactions]*, TAA-GS-05.

[10.5] BG Chemie, *Anlagensicherheit: Exotherme chemische Reaktionen – Grundlagen [Plant safety: Exothermic chemical reactions – Basics]*, Instruction leaflet R001, 11/95, ZH 1/89.

[10.6] KLETZ, T. A., *Cheaper, Safer Plants – Notes on Inherently Safer and Simpler Plants*, 2nd ed., Institute of Chemical Engineers, Rugby, **1985**.

[10.7] BG Chemie, *Accident prevention instruction 64 (VBG 64), Wärmeübertragungsanlagen mit organischen Wärmeträgern [Heat-exchange units with organic heat-transfer media]*, October **1993**.

[10.8] BG Chemie, *Richtlinien für die Vermeidung der Gefahren durch explosionsfähige Atmosphäre mit Beispielsammlung – Explosionsschutz-Richtlinien – (EX-RL) [Guidelines for avoiding the dangers of explosive atmospheres with a collection of examples – Guidelines on protection against explosions – (EX-RL)]*, 6/96, ZH 1/10.

[10.9] Prescriptions on electrical units in explosive areas (ElexV). Issue no 96, Federal Law Gazette (of Germany).

[10.10] Technischer Ausschuss für Anlagensicherheit (TAA) [Technical Committee for Plant Safety], *Rückhaltung von gefährlichen Stoffen aus Druckentlastungseinrichtungen [Containment of hazardous dangerous materials from pressure-relieving installations]*, TAA-GS-06.

11
Choice and Optimization of Multiproduct Plants

11.1
Introduction

This final chapter will deal with the most important factors determining the choice and optimization of the different versions of multiproduct plants. Because existing plant structures are in many cases included in the planning of a multiproduct plant – the "blank slate" as starting point being more the exception than the rule – close attention will be paid to the upgrading of existing plants to multiproduct plants.

In multiproduct plants, the process technology is usually predetermined to some extent. That is, each of the products manufactured in such a plant is produced by the same sequence of process steps, or the sequence of process steps is already determined by the class of the products. In plants with multipurpose equipment, in contrast, it is already sufficient when the desired basic operation is available. In this context, process-engineering-type considerations play a less important role.

Five basic characteristic functions of multiproduct plants can be inferred from this:

- Developing new process concepts (a function less important in multiproduct plants than in monoplants)
- The choice of possible plant concepts or types
- The upgrading of existing monoplants to multiproduct plants
- The revision or optimization of applied processes in multiproduct plants
- The integration of new products in existing multiproduct plants

The function of *developing new process concepts* is that of determining process steps and suitable equipment. The process steps and equipment types can be chosen and combined freely. The objective is to create a flexible solution, making process variations economically viable. In multiproduct plants, process synthesis is generally not the predominant issue. Optimizing the production process of a single product can also not play a major role in a multiproduct plant. It will always be a compromise, because the production cannot be optimized for each individual product; rather, it needs to be optimized for a more or less large number of products in the same plant, and thus each product will be produced by a method that is not completely optimal (product-assortment flexibility).

For the *choice of possible plant concepts or types*, location-specific and product- or product-class-specific requirements play the dominant role. The process-technological interests are generally already predetermined. Important, once again, is the technology, fixed by the choice of the specific type of plant. For *upgrading existing monoplants to multiproduct plants*, the same principles apply; once again the technology associated with the monoplant is predetermined.

For the *optimization of an already existing process* or the *integration of new products into an existing multiproduct plant*, the process steps and equipment are generally predetermined. Only minor modifications are usually possible, such as those limited to removing bottlenecks, and modification and integration of individually available, rapidly accessible, or new equipment.

For optimization, the conditions used as starting point differ from those used for process development. Particularly in multiproduct plants, influences such as changes in product portfolio could have changed the marginal conditions so frequently since commissioning of the plant, that the nature of the task of the current production needs to be formulated completely anew. It may even be possible that the originally planned plant can no longer operate optimally under the prevailing conditions.

Technical progress in the form of improved technology and equipment and additionally developed methods and tools for designing plants and for process control are further inducements. In addition, as a result of more intense competition and shorter product cycles, the question of whether it is possible to use existing plants also for new products is nowadays being asked much sooner than before.

The first question that needs to be answered when a new plant is to be built is whether the plant is to be a single- or multiproduct plant.

New plants that are to be operated as development units – for the development of processes and/or production of market-introduction quantities – will most probably be conceived as multiproduct plants. Such processes are generally only run temporarily in the plant, because of the short time span of such developing products in pilot plants, so that the inevitable next question for the plant will be: "What next?"

It is therefore crucial that such plants *and their equipment* are from the start designed so that a later function, which may be unknown at the time of planning, is possible after only minor alterations.

For production plants, it is more difficult to answer the question.

The operational earnings of a production plant is determined as the difference between the product costs and the sales, depending on the quantities sold (Fig. 11.1). In dedicated plants, that is, plants optimally adapted to a process or product, the individual cost types can be minimized. However, when small quantities are produced in dedicated plants, the capital costs of the plant play an increasingly important role, so that its trend is in the opposite direction to the variable costs, which depend on the efficient use of the capacity to which the plant is operated. The increasing influence of a plant's capital costs on the total production

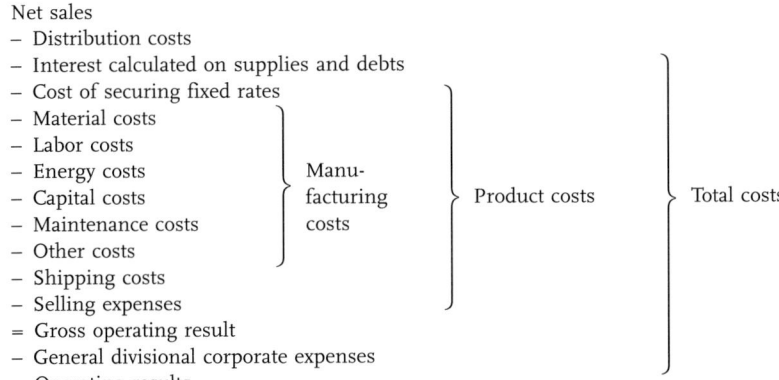

Fig. 11.1 Cost structure of a production plant

$$\frac{I_1}{I_2} = \left(\frac{Q_1}{Q_2}\right)^a$$

I   Investment costs
Q   Plant capacity
a Regression exponent, 0.5–0.9

Fig. 11.2 Calculation of investment costs by regression

costs of a product with decreasing plant size is given by the exponent of the regression curve describing the investment costs of plants of various sizes (Fig. 11.2). The exponents are between 0.5 and 0.9, which means that, depending on the plant type, the specific investment costs per kg product increase with decreasing plant size [11.1].

It should also be kept in mind that labor costs are not proportional to the size of the plant; for plants below a threshold plant size, these costs tend asymptotically to a constant value. This means, for example, that shift operation can only be maintained with a minimum number of personnel, so that reduction in the number of staff is no longer possible. All of this means that, in the end, the specific costs of products produced in small plants are higher.

At which plant size do the advantages of a plant adapted specifically to the process and the product counterbalance the higher specific costs of that plant? To this question there is no general answer; this issue has to be weighed up for each plant individually. For products where the capital costs make up a big proportion of the total production costs, and in plants that do not run to sufficient capacity, it should be checked whether the profitability of production cannot be improved by the realization of a multiproduct plant concept to improve the use of the capacity of the plant.

11.2
Selection of Multiproduct Plant Type

11.2.1
Evaluation of Plant Concepts by Flexibility Criteria

The different types of flexibility of process-engineering systems were already introduced in Section 1.2:

- Structural flexibility: the type of flexibility that allows a system, through changes in the connections between its elements, to adapt to changed demands in function.
- Product-assortment flexibility: this is the ability of a system to produce different products without it being necessary to change the system substantially.
- Flexibility in capacity: the property of a system that can accommodate different capacity demands.

A careful analysis to determine the most important type of flexibility requirements should precede the choice of a specific type of multiproduct plant [11.2]. A comparison of the different types of multiproduct plants show that they differ substantially with regard to flexibility (Tab. 11.1). Continuously operated standard multiproduct plants and plants with multipurpose equipment have not been included in this comparison, since other criteria play important roles in these plant types.

The high structural flexibility of modular multiproduct plants is due to the mobility of the peripheral units. In multiproduct plants with pipeline manifolds, the main equipment and the periphery are both fixed. The high structural flexibility, in this case, is achieved through the multiple connection possibilities of the piping. With pipeless plants, the periphery (the functional stations) is fixed, while the main equipment is mobile. This results in this plant type having low structural

Tab. 11.1 Comparison of different types of multiproduct plants with regard to flexibility

Plant type	Main equipment	Periphery	Structural flexibility	Product-assortment flexibility	Flexibility of capacity
Standard multiproduct plant	fixed	fixed	low	high	low
Modular multiproduct plant	fixed	mobile	high	medium	low
Multiproduct plant with pipeline manifolds	fixed	fixed	high	medium	high
Pipeless plant	mobile	fixed	low	low	high
Monoplant	fixed	fixed	low	low	low

flexibility, as only a few types of functional stations, the charging-, filling-, and mixing stations, are available at any one time.

Product-assortment flexibility is basically determined by the variety of available operating conditions. The standard multiproduct plant with its fixed piping configuration can offer the biggest range of operating conditions. Modular multiproduct plants and multiproduct plants with pipeline manifolds have a limited range of operating conditions available, because of the connections between the equipment that need to be opened and shut. The product-assortment flexibility of pipeless plants is low, because only open mobile containers could so far be used in this plant type.

The flexibility in capacity of standard multiproduct plants, modular multiproduct plants, and plants with multipurpose equipment is relatively low, as the capacity of all these plant types is determined by the fixed size of the reactor. In contrast, variously sized items of main equipment can be used in multiproduct plants with pipeline manifolds and pipeless plants.

In practice, mixtures of these plant types are often found, for example, a standard multiproduct plant combined with elements of a modular multiproduct plant and pipeline manifolds.

11.2.2
Selection of a Suitable Multiproduct Plant Concept

The planner of a chemical plant will, as a rule, have further marginal conditions to consider. The options will be limited further by the following parameters (see Tab. 11.2):

- Material properties
- Operating conditions
- Production functions
- Space considerations

If the demands are increased by the toxicity of the substances handled in the plant, then permanently installed plants with fixed piping are preferable; compartmentalized modular multiproduct plants are also an option.

If the substances are difficult to transport by pumping, the pipeless plant concept is of advantage. Demanding operating conditions are realized best in standard multiproduct plants; pipeless plants are still unsuitable for such applications.

If a large number of substances and recipes are used during production, the standard multiproduct plant is inappropriate; the pipeless plant concept has many advantages under these conditions.

Pipeless plants are very well suited to automation, as are recipe-controlled discontinuously operated standard multiproduct plants. As far as we know, there have so far been only a few rare cases of completely automated modular plants or plants with pipeline manifolds. Versions of automated pipeline manifolds are, however, commercially available. With the development of pipeless plants, easy cleaning of the plant, achieved through special cleaning stations, is the main issue.

Tab. 11.2 Matrix for selection of suitable plant type

Plant concepts	Requirements														
	Material properties		Operating conditions			Production functions						Space considerations		
	Toxic substances	Transportation difficulties	Broad temperature range	High pressure	Exothermic reactions	Many substances	Many recipes	Automation possibility	Easy cleaning	Process tech. complexity	Use of special apparatus	Large apparatus	Limited space	Different apparatus heights
Standard multiproduct plant	+	○	+	+	+	–	–	+	○	○	+	+	+	+
Modular multiproduct plant	+	–	○	○	○	○	○	○	○	+	○	–	○	○
Multiproduct plant with pipeline manifolds	○	○	○	+	+	○	○	○	○	+	+	+	+	+
Pipeless plant	–	+	–	–	–	+	+	+	+	–	–	○	–	–

+ very suitable, ○ limited suitability, – not suitable

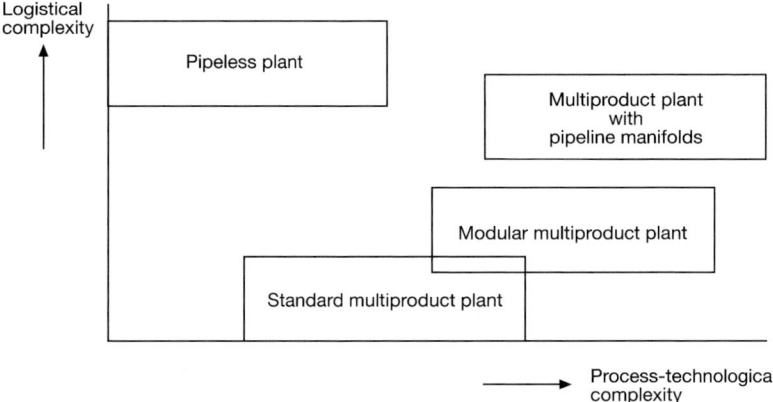

Fig. 11.3 Classification of the plant types according to the logistical and process-technological complexity of the task

Tasks of process-technologically complex nature are only economically realizable in modular plants or plants with pipeline manifolds. Special equipment can be integrated best into standard multiproduct plants and over pipeline manifolds.

In most of the cases, the planner will be confronted with a given space situation. Standard multiproduct plants as well as connections over pipeline manifolds are easily integrated into pre-existing structures. Plants with movable modular units need adequate floor space around the main equipment, and the size of the equipment is also limited by transport facilities such as elevators.

Large horizontal surfaces are required for the transport systems used in pipeless plants. For this reason, this concept is not suitable for situations where an existing structure needs to be converted.

In summary, for multiproduct plants, the logistical components can be as important as the process-technological aspects. For specific production functions, in the coatings sector, for example, the handling of many substances and recipes as well as the diverse transport tasks within the production plant may play the dominant role. If the logistical complexity is superimposed onto the process-technological complexity of a task, the four types of multiproduct plants can be categorized as shown in Fig. 11.3.

Pipeless plants have clear advantages for processes involving extensive and difficult handling of substances and easy process-technological tasks. Complicated process-technological tasks are best dealt with in modular multiproduct plants and plants with pipeline manifolds. Standard multiproduct plants are suitable for process-technological tasks of medium complexity.

11.2.3
Extending a Monoplant to a Multiproduct Plant

Quite a number of multiproduct plants have been developed from monoplants. This could be because production of the originally intended product has come to an end or the sales of the product have not met expectations and the plant could not run to full capacity. Deliberations on how to make better use of the existing plant then often lead to the decision to use the plant for other products by upgrading it with more machinery and apparatus [11.3].

Before such considerations can be put into practice, various marginal conditions need to be addressed first. Even unfavorable marginal conditions can, of course, be overcome with sufficient investment, but the economic pressure is usually strongest precisely where existing plants are upgraded into multiproduct plants, leaving very little financial maneuvering space. The usual pay-out calculations are very useful for an evaluation prior to decision-making.

Spatial Considerations
If very little space is available, standard multiproduct plants, plants with multipurpose apparatus, and multiproduct plants with pipeline manifolds are the most suitable options.

A prerequisite for the efficient handling of solids is the possibility of handling that product in only one container or effecting material flow over several levels, usually five to six levels: (1) Handling of feedstocks, (2) Agitated reactor vessel, (3) Buffer containers/Agitated reactor vessel, (4) Centrifuge, (5) Shovel drier, and (6) Packing/filling.

Apart from the plant itself, the space required for peripheral units, such as secondary heating/cooling circulation (see Section 3.7), storage tanks, and logistics, should also be accounted for.

Infrastructure Facilities
Multiproduct plants place higher demands not only on the plant technology, but also on the infrastructure.

Systems for waste-water and waste-gas disposal need to be universally designed; commonly used in multiproduct plants are, for example, scrubbers with high-temperature combustion at 1200 °C, with defined residence times and subsequent flue-gas scrubbers. Various temperature levels to control the temperature of the plant, as well as low-temperature refrigerating units are standard equipment in multiproduct plants.

If such systems need to be added for an existing dedicated plant to be upgraded to a multiproduct plant, substantial financial investment, which may approach the value of the plant itself, is required [11.4].

Plant

This issue is deliberately addressed at the end of this section, and forms the transition to the next part, as improvements to a plant itself are usually possible with some expenditure.

Particular attention should be paid to the materials used as well as the possible pressure and temperature conditions in the plant. Changes to these parameters are not practical or are equivalent to building a new plant.

11.3
Process-Engineering Optimization of Multiproduct Plants

11.3.1
Approach

The usual approach for developing, revising, and modifying multiproduct plant concepts, apart from the tools appropriate for the individual phases, is shown in Fig. 11.4 (see also references [11.5, 11.6]). The four main phases are:

- Analyzing the task and considering various concepts (new plant) or taking stock and analyzing the existing concept (pre-existing plant)
- Investigating different possibilities and choosing and checking alternatives
- Developing the details of the chosen alternative
- Evaluating the risks, safeguarding, and implementation

The nature of the task is first analyzed and then a heuristically- and experience-based *process concept* is worked out as starting point for cost- and structural consid-

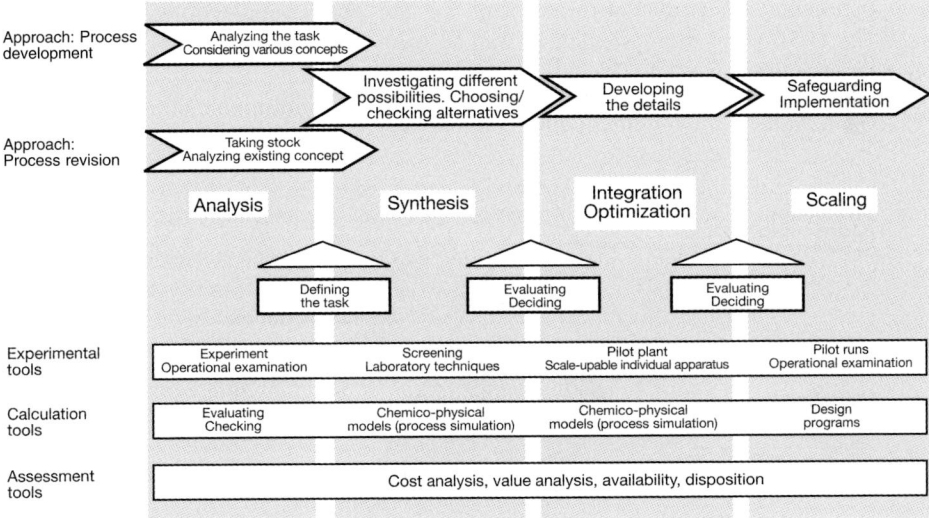

Fig. 11.4 Approach and tools

erations. The goals and their time- and cost frames are then determined. In multi-product plants, this phase is often of minor importance.

For the optimization of a process, as with integrating new products, the first step is that of *taking stock*. The objectives are usually cost reduction, implementation of general guidelines, or the economic production of new products with the equipment already available. For this, the cost structure is analyzed according to the type and nature of the task. The final result is a decision on which themes to deal with in which order, or which objective guidelines make the most sense.

This stage is followed by the *investigation*, development, choosing, and checking of different alternatives, which are compared and evaluated, leading to the most promising possibility being selected.

After the *details of the chosen alternative* have been *developed*, the possible solution is evaluated and it is decided whether the concept is viable and whether the accompanying risks are acceptable. If necessary, another alternative needs to be developed.

The next step is to address the *implementation* of the design; the equipment needs to be planned according to what is required, possible remaining risks need to be safeguarded against, and the solution developed thus is implemented.

Fig. 11.4 represents the structure of this procedure, in which individual phases may overlap. It is important that the actual problem is defined, evaluated, and decided upon. This is often an iterative process: many of the phases may be repeated, but each time the state of knowledge is different. A new route is therefore the result of additional information and/or changed emphasis.

The development of new processes and the optimization of existing ones (or the integration of new products into existing processes) only differ in their first phases. The content of the next three phases is very similar regarding procedures and tools.

For multiproduct plants, sound knowledge of the products and process sequences, possibly supported by process simulation or laboratory- or operational experiments are particularly necessary for the process-engineering analysis (see Fig. 11.5). From this systematized and abstracted knowledge, the operating characteristics are compiled. These form the basis of an economically viable process operation; as a basis, they serve the new design or optimization of a multiproduct

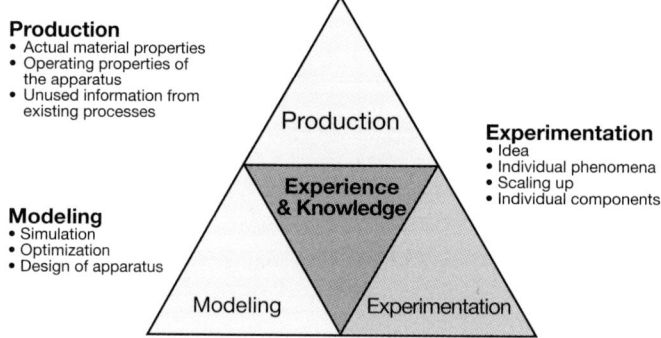

Production
• Actual material properties
• Operating properties of the apparatus
• Unused information from existing processes

Experimentation
• Idea
• Individual phenomena
• Scaling up
• Individual components

Modeling
• Simulation
• Optimization
• Design of apparatus

Fig. 11.5 The elements required for process development

plant and the integration of new products (besides the effects on the product palette) equally well.

11.3.2
Information from Existing Processes

For optimizing an existing process, as well as for integrating new products into an existing plant, important data are obtained from the production process already in operation. It is important to use, classify, and evaluate this available information.

This information on the process sequences in *production* plants (see Fig. 11.6) is often unutilized, because it is not "obvious," that is, difficult to gain access to, or not necessary, for production. It is, as a rule, possible to balance the quantity and material flow in the actual plants (operational data). In individual cases, for obtaining a complete economic analysis, it may be useful to carry out experiments in a production plant. However, the preconditions are always that production should not be hindered, the safety of the operation is guaranteed, and the quality of the products remain intact (as far as possible). This process ensures that one is dealing with real product- and plant conditions.

11.3.3
Knowledge and Experience

Knowledge and experience make up a substantial part of the results obtained by experimentation and modeling and the application of these to the real situation of a new or modified multiproduct plant. The next part will therefore focus on a few aspects important to practical situations; evaluating and process-relevant tools will also be dealt with briefly.

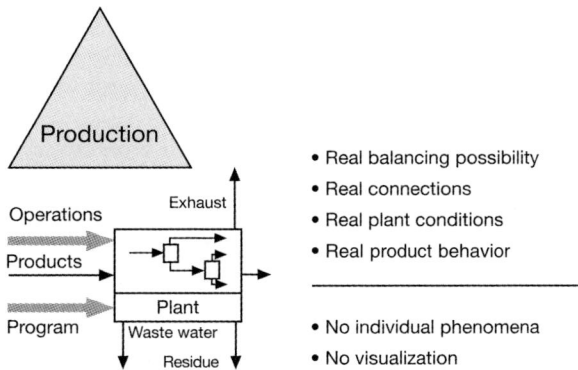

Fig. 11.6 Potential and limits of experiments and examinations in operating plants

Surrounding Field Conditions

Knowledge of the surrounding field conditions of a new plant often plays a decisive role in ensuring realistic process development. The advantages of some sites, due to reverse integration and the availability of raw- and auxiliary materials, supply and disposal facilities, as well as technical and economic know-how, over the "clean-slate" situation, without waste-water treatment facilities, with other infrastructure areas in need of renovation, and where new skills need to be learned for production, are obvious. This explains how important the site conditions can be.

When alternative sites are considered, it is often useful to compare the quality of the formally identical chemical starting materials that are on offer at these different sites, otherwise additional preparation steps for removing particular impurities may end up being essential at the chosen site. Questions such as the arrangement of plant and products as well as logistics also play an important role.

Tools for Evaluation

An essential part of process development is the rapid and reliable evaluation of the newly developed concepts. From the value analysis area, tools based on knowledge and experience are available for this purpose, to give fundamental insights into the actual situation of a process.

The *cost structure* analysis of a total process can, depending on the specific requirements, contain different balancing limits. This may also mean that the components used in the process under consideration may not even be present in another balancing limit, and that they first need to be synthesized by another production process, so that the feedstocks and production costs also need to be taken into consideration. Different balancing limits can easily lead to different economic results.

Every process is ultimately influenced by costs, but it is only the *allocation of the costs* and looking at the types of costs involved in the individual steps of a process that makes it possible to see the potential associated with the individual steps and where activities pay off.

This analysis alone may not be sufficient. A more precise picture is given by the *costs that can be influenced*, that is, approximately the raw-material costs that have gone into the "lost yield," the difference between a 100% yield and the actual yield obtained. The extra expenses found in reality represent the actual improvement potential.

An analysis of the cost structure of a process may not be good enough, as this approach implicitly accepts the given structure of a process and does not question it. Another vantage point becomes possible only if one analyzes the functions of the individual basic operations. Therefore, the methodically correct starting point (with regard to a value analysis) is the analysis of the *functions* of the individual steps. Only such an approach allows other connections and other separation operations to be considered.

Specific Tools for Process Optimization

All the tools and aids of process development, such as screening methods for choosing auxiliary materials, simulation programs for overall processes or individual steps, dimensioning programs for columns, and so forth, are, of course, applied during process optimization. In this section, the process optimization tools used most often and for this specific purpose are focused on.

"*Debottlenecking*" is a task that inevitably crops up in the context of multiproduct plants. Generally, bottlenecks are identified very pragmatically: production is increased until it "gets stuck" somewhere. Only one bottleneck can be identified at a time by this method. The same method has to be applied continuously for the next bottlenecks to be found.

The following method, best applied to processes that are largely calculable, is more systematic. The starting point is a comparison of the actual operating points with the operability areas of the individual apparatus, connected to a thermodynamic–physical simulation of the process. If the relevant operating points as well as the possible operating areas of all the equipment are normalized relative to the throughput based on the product of value, this method is reliable for the identification of bottlenecks and the evaluation of their significance for the process as a whole.

Balancing of running processes is often possible with simple in-house or commercially available programs for balancing quantities and materials, mostly without or with only rudimentary thermodynamic elements. In practice, many and various data are available for describing a stationary operating status; these data differ by slight degrees and are affected by measurement and analysis errors. Furthermore, data are often obtained from more check points than are necessary for consistent balancing. This leads to a mathematically overdetermined system. Until recently, attempts at checking such operational data have been by means of simulation that proceeds until the observer is convinced that the simulation result describes reality. Programs that allow the probable operating status to be determined from these operational data – more independent of the observer and also more objectively than possible so far – have recently become available. These programs also contain almost complete thermodynamic models, so that the calculated solutions are thermodynamically plausible; in addition, they also indicate how the reliability of the result can be improved by further information at specific points in the process [11.7, 11.8]. In other words, the results show where additional measurement points are necessary, where measurements are superfluous, and where available measurement points should be checked or adjusted.

11.4
References

[11.1] PRINZING, P., RÖDL, R., AICHERT, D. *Chem.-Ing.-Tech.* **1985**, *57*, 8–14.

[11.2] FÜRER, S., RAUCH, J., SANDEN, F. *Chem.-Ing.-Tech.* **1996**, *68*, 375–381.

[11.3] SCHUCH, G., KÖNIG, J. *Chem.-Ing.-Tech.* **1992**, *64*, 587–593.

[11.4] THIER, B. *3R International* **1985**, *24*, 648–657.

[11.5] BESSLING, B., CIPRIAN, J., POLT, A., WELKER, R. *Chem.-Ing.-Tech.* **1995**, *67(2)*, 160–165.

[11.6] CIPRIAN, J., POLT, A. *Chem. Technik* **1996**, *48(6)*, 293–300.

[11.7] WENDELER, H., GOEDECKE, R., JANOWSKI, G., KULTAU, W., *Prozeßdatenvalidierung von Chemie-Anlagen*. GVC Fachausschuss „Prozeß- und Anlagentechnik" [Working party on Process- and Plant Technology], Wiesbaden, **1992**.

[11.8] LIST, T., DEMPF, D., *Erfahrung mit Prozeßdaten-Validierungs-Systemen*. GVC Annual Meeting, Dortmund, **1996**.

Subject Index